The Science and Best Practices of Behavioral Safety

This book presents the scientific principles and real-world best practices of behavioral safety, one of the most mature and impactful applications of behavioral science to reduce injuries in industrial workplaces.

The authors review the core principles of behavioral science and their application to modern safety processes. Process components are discussed in detail, including risk analysis and pinpointing, direct observation, performance feedback, reinforcing engagement, trending and functional analysis, behavior change interventions, and program evaluation. Discussions are complemented by industry best-practice case studies from world-class behavioral safety programs accredited by the Cambridge Center for Behavioral Studies (CCBS), which provide compelling evidence of the effectiveness of these behavioral science principles in reducing injury.

The Science and Best Practices of Behavioral Safety is essential reading for safety professionals, process safety engineers, and leaders in companies who have implemented, or are considering implementing, behavioral safety; or as an aid to learning more about the scientific background behind effective and practical safety practices. Researchers, expert consultants, and students who are already familiar with the practice will also find the book a valuable source to further develop their expertise.

Timothy D. Ludwig, Ph.D., has more than 30 years of experience in research and practice in behavioral safety. Dr. Ludwig is a Distinguished Graduate Professor at Appalachian State University; serves on the Cambridge Center for Behavioral Studies (CCBS) Commission for the Accreditation of Behavioral Safety; and disseminates his writings on Safety-Doc.com.

Matthew M. Laske, M.A., received his master's in industrial-organizational psychology from Appalachian State University and is completing his doctoral degree at the University of Kansas. Matthew has designed, implemented, and assessed behavioral safety programs in multiple industries and is recognized by the CCBS as a Distinguished Scholar.

"As a researcher and teacher for more than 50 years, I've seen remarkable reductions in workplace injuries with applied behavioral science. This teaching/learning text reviews the research-based evidence and shows how to apply the principles to make optimal impact."

E. Scott Geller, Ph.D., *Virginia Tech,*
Safety Performance Solutions, USA

"The science is what my company relied on to deliver behavioral safety programs with confidence. This book locks in the principles and evidence that our community has used for the past 40 years to reduce injuries."

Terry McSween, Ph.D., *Quality Safety Edge, USA*

"There are several on the market but this book's details would help actually make changes without consultants. You see what it actually takes to make changes with enough examples and data proving that behavioral science can be implemented in complicated corporate settings."

Angie Lebbon, Ph.D., *Eastman Chemical, USA*

"A thorough review and contemporary analysis of what behavioral safety is and what best practices make it effective in reducing worker injuries. Essential reading for safety professionals and practitioners of applied behavioral science."

Cloyd Hyten, Ph.D., *Director of Safety Solutions, USA*

The Science and Best Practices of Behavioral Safety

THE SOURCE FOR REDUCING INJURIES ON THE FRONT LINE

Timothy D. Ludwig and
Matthew M. Laske

Routledge
Taylor & Francis Group

NEW YORK AND LONDON

Cover Image: © Getty Images

First published 2023
by Routledge
605 Third Avenue, New York, NY 10158

and by Routledge
4 Park Square, Milton Park, Abingdon, Oxon, OX14 4RN

Routledge is an imprint of the Taylor & Francis Group, an informa business

Library of Congress Cataloging-in-Publication Data
Names: Ludwig, Timothy D., author. | Laske, Matthew M., author.
Title: The science and best practices of behavioral safety : the source
    for reducing injuries on the front line / Timothy D. Ludwig,
    Matthew M. Laske.
Description: New York, NY : Routledge, 2023. | Includes
    bibliographical references and index.
Identifiers: LCCN 2022035100 (print) | LCCN 2022035101 (ebook) |
    ISBN 9781032256009 (pbk) | ISBN 9781032269672 (hbk) |
    ISBN 9781003290711 (ebk)
Subjects: LCSH: Behavior modification. | Industrial safety. |
    Psychology, Industrial.
Classification: LCC BF637.B4 L84 2023 (print) | LCC BF637.B4 (ebook) |
    DDC 153.8/5—dc23/eng/20221116
LC record available at https://lccn.loc.gov/2022035100
LC ebook record available at https://lccn.loc.gov/2022035101

ISBN: 978-1-032-26967-2 (hbk)
ISBN: 978-1-032-25600-9 (pbk)
ISBN: 978-1-003-29071-1 (ebk)

DOI: 10.4324/9781003290711

Typeset in Palatino
by Apex CoVantage, LLC

DEDICATION

This book is dedicated to the safety professionals, executives and, most importantly, the dedicated workers who strive daily to make behavioral safety successful in reducing human suffering for their workforce and, through dissemination, the world. We thank all these individuals, who have taught us valuable lessons to help us disseminate their best practices to the world. We would also like to show appreciation to our life partners, Lori and Maira, for their constant support both throughout the development of this book and in life.

PUBLISHER'S NOTE

Portions of this book appear in the following publications with permission of the publisher Taylor & Francis:

"Behavioral Safety: An Efficacious Application of Applied Behavior Analysis to Reduce Human Suffering," by T.D. Ludwig and M.M. Laske, 2022, *Journal of Organizational Behavior Management* (Taylor & Francis, 2022).

Ludwig, T.D. & Laske, M.M. (in press). "The Impact of Organizational Behavior Management in Industrial Safety – Case Studies in Behavioral Safety." Chapter in *The Organizational Behavior Management Handbook* (Johnson & Johnson, eds).

CONTENTS

PREFACE

Behavioral safety is one of the most mature and effective applications to reduce injuries in industrial workplaces. Not everyone gets it right; others have had great success and want to take it to the next level. We saw the need for a guidebook that takes everything we know about behavioral safety and puts it in one place. But not some consultant's proprietary behavior-based safety (or whatever they call it) package in a marketing book based on their sales version of their process. Instead, we need to cut through the marketing and look at the underlying science behind behavioral safety. That's right—*science*.

Behavioral safety, from its beginnings to the present, has been built on the foundation of behavior analysis—otherwise known as the *science of behavior*—started by B.F. Skinner. But sometimes this science can get somewhat complicated and can be hard to decipher. Our goal with this book is to translate the science into bite-size chunks that you can put into practice. Don't worry, we will teach you the science as well: we've included "Science Moments" throughout the book, for those interested, to explain some of the core scientific concepts for your deeper understanding. As you learn about the science of behavior and how it has been applied, you will begin to appreciate that behavioral safety is not just some made-up process that happened to become popular. Instead, you will learn that it has been built meticulously, tested over and over, and is continually improving as we learn new things! So, in a way, this book is a textbook teaching the science behind behavioral safety. The authors of this book, Tim Ludwig and Matt Laske, are behavior scientists with a deep understanding of behavior. We will guide you through this science.

Science without practice, however, is just a bunch of scholars writing esoteric, hard-to-read publications for each other. How do we know the science works in the real world? We have to go and do it; fail; find success; grow that success; learn from both the successes and the failures; and finally go back to the science and test it all over again. We've been doing this since the early 1970s and have a lot to show for it. In 2005, a group of scientists who also have decades of experience in the practice of behavioral safety created the Commission for the Accreditation of Behavioral Safety as part of their work with the not-for-profit Cambridge Center for Behavioral Studies (CCBS). As part of this work, they went around to the very best behavioral safety programs in the world to find out what makes them great. Upon awarding its prestigious accreditation, the Commission documented the programs' best practices and evidence of success. This book presents these best practices in a way that is easy to understand and easy to implement. So, in a way, this book is also a guidebook, leading you through the steps to implement these best practices.

Behavioral safety aims to prevent harm and reduce human suffering by targeting risk and intervening to promote safe behaviors. This book:

- reviews the core components of a behavioral safety process;
- highlights modern behavioral safety methods (e.g., behavioral systems analysis); and
- reviews best practices from world-class behavioral safety programs accredited by the CCBS.

Whether you are just starting your behavioral safety process or want to take it to the next level, we hope the science and practice documented in this book will lead you to success in reducing harm in the workplace.

ACKNOWLEDGMENTS

The authors would like to thank the Cambridge Center for Behavioral Studies (CCBS) (behavior.org), Dwight Harshbarger and Bill Hopkins for the vision to document the best practices of behavioral safety programs through the CCBS Commission on Behavioral Safety Accreditation. We would also like to thank the members of the CCBS Commission—including Siggi Sigurdsson, Angie Lebbon, Oliver Wirth, Mark Alavosious, Don Kernan, Sandy Knott, Alan Cheung, and Distinguished Scholars Andressa Sleiman and Nicholas Matey—for documenting and reviewing the behavioral safety programs whose efficacious practices we present in this book.

1

WHY BEHAVIORAL SAFETY?

Why Behavioral Safety?

A man died as he slid under a shredder to remove a jam. He had neglected to follow the lock-out tag-out process, opting for a quick fix. When he removed the compressed cotton blocking a blade, the machine kicked into operation and he was shredded instantly. He was a supervisor with 30 years of experience. This event shocked a young HR executive and galvanized his lifelong passion for safety. He was trained in applied behavior analysis, had spent some time as a scientist and now was applying his science in the real world of industry and risk. He immediately engaged his company in a worldwide campaign to apply behavioral safety—a method designed and tested by many in his science—to transform safety, quality, and the work lives of the thousands of workers under his care. He did the same in his second executive appointment with a shoe company, whose workers in overseas plants were getting injured at unacceptable rates. Consumers in America and Europe would not put up with reports of unsafe plants causing harm to vast workforces laboring in faraway countries. Investments were made in these plants, the equipment, and facilities; but injuries continued at high rates. Something had to be done. This was a problem across industries around the world. A new approach to safety was to be found in the behaviors of workers and the science studied in applied labs across academia.

It all started as a sort of movement among researchers who attempted to translate the science of behavior analysis into real-world applications to better society. These were scientists who studied human behavior. B.F. Skinner was still an active thought leader in the early 1980s and began to imagine utopias where the science of behavior could be applied to relieve human suffering and bring out the best in people. Behavioral safety is firmly rooted in the scientific field called "applied behavior analysis." Behavior analytic terms and principles can be quite esoteric (meaning that few people can understand what they mean). Therefore, we have inserted "Science Moments" throughout the book. Science Moments are flagged in the text like this: [SM] In these Science Moments, we will briefly explain the scientific terms and principles behind behavioral safety. This way, you can better understand the integrity and precise operations behind human behavior. Our first Science Moment provides a brief history of applied behavior analysis[SM-1.1].

DOI: 10.4324/9781003290711-1

Science Moment 1.1
A Brief History of Applied Behavior Analysis

Applied behavior analysis is the application of the principles of behavioral science to socially significant behavior. The roots of applied behavior analysis are often attributed to the experimental work done by B.F. Skinner (Morris et al., 2005). Skinner was interested in what variables influence behavior. Although Skinner primarily worked with pigeons and rats, he knew the principles could be applied to humans. Skinner's *Science and Human Behavior* (1953) described how behavior analysis could be applied to socially important human behavior. What came next was an explosion of applied behavioral research; behavioral applications to workplace behavior were not far behind and increased dramatically. Applications to workplace safety were quick to follow, with Beth Sulzer-Azaroff publishing the first application of behavioral principles to workplace safety in 1978.

It was a challenge to take a science whose roots were in animal research, expand its principles through human research, and then find efficacious ways to apply the science to real-world problems with evidence and best practices. A legion of professors, students, and consultants began to test behavioral principles with great success in industry and business. When the world began calling for a better way to protect frontline workers from harm in hazardous manufacturing, petrochemical, distribution, mining and construction sites around the world, the science of behavior answered the call.

Dwight Harshbarger had studied behavior since his undergraduate days at West Virginia University. His studies took him to Berkeley, the University of North Dakota and Harvard before an appointment with the faculty of West Virginia University. Dwight had a keen need to help people, grounded in his rural upbringing. He witnessed the plight of families of workers injured in West Virginia's mines and began to document how some of the world's largest companies were putting their workers and communities at risk. Thus, he ventured away from academia to lead a community mental health center in southern West Virginia. Now he set his sights on applying his science to safety, first as a consultant and then as an executive within the corporate world responsible for the safety of international workforces.

Galvanized by the fatality under his watch, Dwight called on his good friend Beth Sulzer-Azaroff, a top researcher at the University of Massachusetts Amherst breaking new ground in behavioral applications to safety. She was following up on the work of Judith Komaki (Komaki et al., 1978) and Kent Anger (M.J. Smith et al., 1978)—both giants in the field—targeting behaviors related to safe work performance through direct observation and analyses (Fellner & Sulzer-Azaroff, 1984; Sulzer-Azaroff & Fellner, 1984; Sulzer-Azaroff, 1978). Dwight took Beth to one of his most challenged plants in Thailand and engaged the workforce in one of the many initial behavioral safety programs that were fast becoming a global phenomenon.

Behavioral safety was quickly developed, tested, and deployed by a peerless generation of behavior scientists into evidence-based programs marketed to industries eager to reduce injury rates under the moniker "behavior-based safety" (Agnew & Snyder, 2008; Geller, 2001a; Krause, 1997). Behavioral safety programs like the one Beth and Dwight implemented back in the 1970s are still going strong some 50 years later because of their demonstrated success in reducing injuries. A lot has been discovered in the science of behavior over these past 50 years and much has been learned in the practice of behavioral safety. Fortunately, the discipline of documenting the science has resulted in hundreds of published research articles validating different components of behavioral safety. Professional conferences and journals have allowed experienced practitioners to share their translation of the science and best practices discovered through implementing behavioral safety in real-world scenarios.

Dwight went on to become the executive director of the Cambridge Center for Behavioral Studies (CCBS) (behavior.org). The CCBS is a not-for-profit organization founded in 1981 whose mission is "to advance the scientific study of behavior and its humane application to the solution of practical problems, including the prevention and relief of human suffering." In 2005, Dwight teamed up with Bill Hopkins—himself a top researcher in behavioral safety—to establish the CCBS Commission on Behavioral Safety, which sought to accredit the world's best-in-practice behavioral safety programs. Since Dwight and Bill's first accreditation of the behavioral safety program of Eastman Chemical's Acetate Fibers Division, the CCBS has accredited over 30 behavioral safety programs, verifying that their processes are based on behavior science and have been proven effective by significant reductions in injuries (under industry standards) for at least three years after implementation of the program. This book presents the science behind behavioral safety and celebrates the programs that have translated the theory into best practices.

Let's begin by understanding why behavioral safety is still as important today as it was 50 years ago, when it was first applied from the science. It's all about the people. Workplace injuries have serious personal costs. The Bureau of Labor Statistics (2020ab) reported 5,333 work-related deaths and 888,220 lost-time injuries in 2019 in the United States. These are real people whose lives have changed just because they went to work to support their families. Pain and disability due to injury cause substantial limitations in activities such as exercise, household chores, and family interactions (Dembe, 2001; Strunin & Boden, 2004). In an estimated 40% of cases, a family member had to reduce time committed to their household, work, and/or schooling activities to replace contributions truncated by the injured worker's limitations (Hensler et al., 1991). Getting hurt can be expensive too. Lifetime costs for a person suffering a nondisabling workplace injury have been estimated to average $10,000, rising to $30,000 for injury-caused disabilities (Marquis & Manning, 1999). This is in addition to lost earnings ranging between $42,100 and $68,100 (Reville et al., 2001); workers' compensation benefits replace only between 32% and 41% of 10-year pretax losses.

In addition to the humanitarian and social implications of severe injury and death, many organizations experience increasing monetary costs related to injuries in the workplace. The Liberty Mutual Workplace Safety Index (2021) has estimated the U.S. cost of injuries to be $58.61 billion annually. Most of these are direct costs, such as when injured employees' medical expenses are billed to the employer. Indirect costs—such as administrative time costs, increased insurance rates, lost production time, and damaged reputations—may be four times the price of direct costs (Safety Management Group, n.d.). The National Safety Council (2021) estimated the combined total of direct and indirect costs to be $171 billion in 2019. The estimated cost for a death averages $1.22 million, whereas that for a medically consulted injury averages $42,000. These costs may escalate for uninsured employees.

Fortunately, over the decades that behavioral safety has been applied, there has been a notable decline in personal occupational injuries in the United States and abroad (Cooper, 2019; McSween & Moran, 2017). We are making progress based on the hard work of safety professionals, consultants, and researchers as we engage in processes that make an empirical difference in safety outcomes like injuries, as well as cultural outcomes like employee engagement in safety and voluntary reporting. That progress includes contributions from behavioral safety programs.

This is why the CCBS accreditation compares the reductions in injuries at sites with behavioral safety programs with the industry average. This first figure (Figure 1.1) shows the original data from Eastman's Acetate Fibers Division showing its injury reduction in comparison to the industry average. Most industries have seen their injuries decrease over the same period, reflecting the impact of other modern safety management systems implemented by the professional safety community. We investigated the impact of behavioral safety programs *over and above* these other systems to find the best evidence-based practices to share in this book.

We hope to convince you that behavioral safety is not just a safety management fad that represented a marketing opportunity for consultants. Instead, we want to reveal the science behind the different components of behavioral safety to show it is rooted in decades of laboratory and field research in applied behavior analysis. We want to show you how the science has been successfully translated into real-life practices in our accredited sites, with real-life results. Finally, and most importantly, we want you to learn about these best practices to improve your behavioral approach to safety.

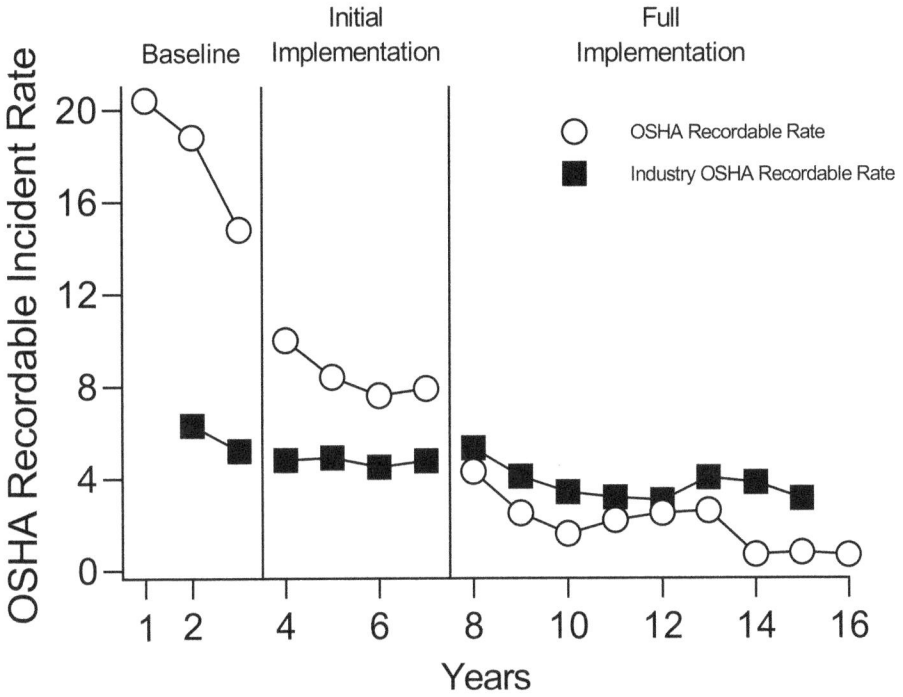

Figure 1.1 Eastman's Acetate Fibers Division Recordable Rate and Industry Standard

Occupational Health and Safety Administration (OSHA) recordable incident rate on the y-axis and consecutive years on the x-axis. White circles indicate Acetate Fiber's OSHA recordable rate. Black squares indicate the industry standard OSHA recordable rate. Phase lines indicate baseline, initial behavioral safety implementation, and full behavioral safety implementation. Data are adapted with permission from CCBS.

THE LIMITATIONS OF SAFETY MANAGEMENT SYSTEMS

We have really made great strides in managing safety in our industries. Many safety improvements over the past decades can be attributed to maturing safety management systems such as the policies, procedures, and activities that promote and maintain the safety of our workforce (Vinodkumar & Bhasi, 2010). The analysis of injuries is the bedrock of most safety management systems. When safety incidents are effectively analyzed, events surrounding injuries can be identified and the causes of the incident mitigated to avoid future injuries. Incident analyses can provide valuable insights into what went wrong and how to fix it.

Thankfully, injuries don't happen very often—especially the types of incidents that cause serious injuries and fatalities. While this is a blessing, it is also a hindrance when using incident investigations to make things safer for your workforce. The problem is that these analyses track low-frequency events (Geller, 1996). Without large numbers, incident analyses are often acting on one-time "special cause variation[SM-1.2]" (Deming, 1982, 1986), which may not prevent future incidents. Further, injury analysis focuses solely on

Science Moment 1.2
Common Cause versus Special Cause Variation

Behavior varies over time; as can the performance that comes out of behavior. Common cause variation is when performance varies but is not drastically different from what occurs on average. If behavior tends to vary consistently (e.g., rate), then the behavior is stable and we can predict how much variability we would routinely expect to see. When exceptional variation is observed in the behavior, different from what is normally observed, this is special cause variation. For example, Figure 1.2 displays the percentage of safe behaviors observed over consecutive weeks of observations. After observing for an extended period, we would have a good understanding of how much variability we should expect in the targeted behavior. Data points in the gray shaded area indicate the common cause variation we might expect. Data points outside of the shaded area indicate special cause variation outside of what we would predict would naturally occur. When we see special cause variation, something significant has changed and has had a drastic effect on performance.

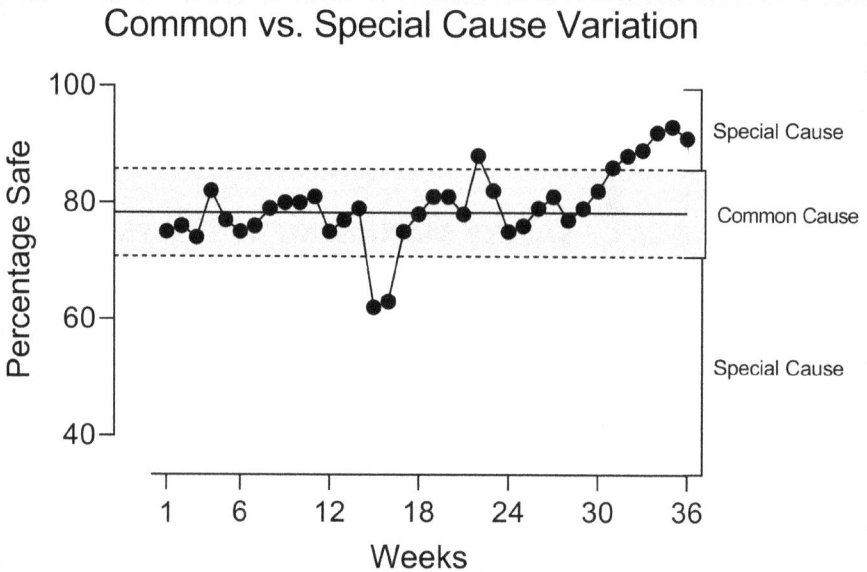

Figure 1.2   Common Cause versus Special Cause Variation

Percentage safe behaviors observed on the y-axis and consecutive weeks on the x-axis. Black circles indicate percentage of safe behaviors observed in a week.

outcome measures responding to events *after* they occur and not prevention *before* they occur (Agnew & Daniels, 2010).

Ideally, the goal of safety management is to identify and mitigate risk before there is an injury. Therefore, reporting of minor injuries and other "leading indicators" (e.g., close calls, equipment damage) is encouraged by employers. These leading indicators have higher base rates that we can use to assess trends in the data. Trends reveal where there is variation in processes, procedures, and—most importantly—human behavior. Human variation is the *real* source of risk. When different people do different things at different times, there is variation. When the same person does different things at different times (e.g., due to fatigue or complacency), there is variation. When people do different things than what are prescribed in the operating procedures, there is variation. Where there is variation[SM-1.3], there is risk.

Trends help reveal the causes of variation (Deming, 1982, 1986) to more accurately describe where injury risk is likely in the work environment. We can analyze the variation we found to forecast more serious injuries. Trends are analyzed and new processes, maintenance, or instructions are applied to mitigate risk (McSween, 1995; Sulzer-Azaroff & Austin, 2000). When we find trends in human behavior, we can analyze the variance to discover why our workforce is at risk. With this understanding of human variation, we can intervene more effectively and efficiently while adapting our safety management systems for more sustainable impact.

Unfortunately, most programs only target the outcome of behavior (i.e., injury, minor incidents, close calls, equipment damage), and not the specific behaviors that put workers at risk while doing their tasks. In many important ways, "leading indicators" are really "lagging" the variance in behaviors that led up to the minor incident, close call, or other event. Think of the injury as the eight-ball in billiards going into the corner pocket (when we played, if you did this accidentally before the final shot, you lost). When we investigate our leading indicators, we try to understand what things got set in motion (variance) that may cause a more serious injury. Thus, we can think of the causes of leading indicators (from our investigations) as the cue ball (the white ball) that hits the eight-ball. All good enough; but what propelled the cue stick forward to strike the cue ball, putting it in motion toward the eight-ball? That would be a human's behavior. We need to understand behavior variation.

The lagging nature of most safety measurements leads to reactive interventions that too often force cumbersome new policies, procedures, and personal protective equipment on the worker (making them less likely to do them), and/or insult their intelligence with retraining and coaching. In contrast, behavioral safety is a preventative scientific approach to safety focusing on changing behaviors before incidents happen.

Science Moment 1.3
Variation

Understanding variation is key to understanding behavior. When there is less variation, we consider the behavior stable. When there is more variation in behavior, we have risk and we must look to the causes of variability. For example, Figure 1.3 illustrates the safe lifting behavior of two people over an eight-hour work shift. Person A has less variability in their behavior during the shift, whereas person B has much more variability in their safe lifting behavior—likely resulting in greater risk. As such, our role is to identify what causes the variability in person B's behavior. Is safe lifting only occurring when the manager is present? Are certain items awkward to lift safely? Are some items too heavy? Is person B more fatigued at different times of the day? Identifying and controlling relevant variables is key to reducing variability and preventing risk from occurring. Person A's behavior is consistent but may not be as high as needed to be safe. When we see consistent performance that is lower than our goal, this represents the current "capability" of the behavior. The worker is doing their best, but this is the best they can do under the circumstances. We need to change their circumstances (i.e., the environment) to impact behavior further.

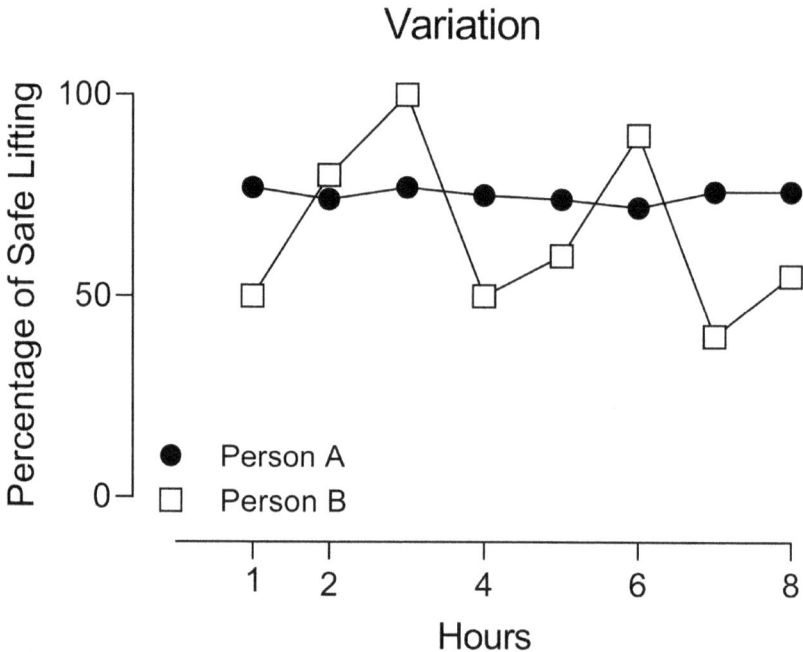

Figure 1.3 Variation

Percentage of safe lifting behavior on the y-axis and consecutive hours on the x-axis. Black circles indicate safe lifting performance for person A. White squares indicate safe lifting performance for person B.

Injury and leading indicator analyses, without the aid of behavioral science, often fail to pinpoint behavioral variance sufficiently. They often only cite rule violations, suggesting variation of process or procedure (or common sense) instead of the actual behaviors within their environmental contingencies. Insufficient analyses inordinately blame the worker for the injury or event.

The link between injuries and at-risk behavior has been widely examined. Reber and Wallin (1983) published one of the first studies reporting a significant correlation between at-risk behaviors and recordable injury rates. McSween (1995) cites one company's internal injury analysis that discovered behaviors as a primary cause of 80% to 90% of injuries. Myers et al. (2010) conducted a similar analysis and found behaviors to be the leading cause of 96% of injuries. Because of this obvious association between behaviors and injuries, investigations too often ascribe the root cause of injuries to "human error" and focus solely on the inadequacies of the employee, as if they intended not to comply with the work process (Cervone et al., 1991). We end up trying to change the person via coaching and retraining, which ultimately is insufficient and often temporary. Workers will go back to behaving the way the environment is set up for them to behave—as will all the other workers who do the job. You tried to change one person when the system you set up maintains the at-risk behaviors of your entire workforce (Ludwig, 2018). Because of this, investigations can fail to adequately engage in analysis of this behavioral variance, understand why it is happening, identify gaps, and implement systems that produce the sustained behavior change that will result in injury reduction.

Because of the predominant focus on injury analysis and the faulty analysis of worker intentions, safety programs have traditionally relied on the use of punishment for rule violations related to injuries. Additionally, leaders try to bribe workers into being safe by offering substantial rewards for an absence of injuries. As a result, these incentive and disincentive programs punish employees' reporting of their injuries and other safety-related incidents (Geller, 1996, 2005a; McSween, 1995). By overusing aversive tactics as a primary safety management system, employee behavior is shaped to avoid reporting incidents (even minor ones). In addition, the managers and safety processes can function as signals[SM-1.4] for punishment (Ludwig, 2018; McSween, 1995). Similarly, incentives offered for achieving a certain number of days without an injury punish injury reporting (Agnew & Daniels, 2010; McSween, 1995). Think of this from the workers' perspective: "If I report an injury, then I have just voided the group reward. Not only do I miss out on the reward, but now I've angered by managers and peers because I've made them lose theirs as well. So I'll just stay quiet with my 'bloody pocket' hiding my injury."

Science Moment 1.4
Discriminative Stimuli

A discriminative stimulus (S^D) is a stimulus that signals reinforcement (or punishment) is coming. In other words, an S^D signals the availability of reinforcement (SR) for a certain response (R). An S^D tends to evoke the behavior consistently. For example, consider a supervisor who only goes into the operating area and gives employees praise for doing inspections. We could diagram this contingency as follows:

S^D (supervisor's presence): R (inspecting equipment) $\rightarrow$ SR (praise from supervisor)

The S^D (supervisor's presence) signals that praise will be available if you start inspecting your equipment. This also means that the supervisor's praise for inspecting is only available when the supervisor is present. If the supervisor is not around, neither is the S^D. In other words, if the workers do inspections in the absence of the supervisor, their behavior will not be praised, leading to extinction of inspection behaviors (i.e., the workers stop doing them).

When the behavior reliably occurs in the presence of the S^D and not in its absence, the behavior is said to be under stimulus control.

These biased analyses and misapplication of rewards and punishment are not what behavior science teaches us. Instead, consider Skinner's (1974) dictum that behaviors are a product of the environment we put the worker in and the systems that control those environments[SM-1.5]. Behavioral analyses attempt to find the cause of the behavior related to injuries instead of the cause of the injury. We certainly do NOT blame the worker; that goes against what our science preaches. Therefore, it is quite unfortunate when those not acquainted with the science condemn behavioral safety as a tool that management uses to blame the worker (Howe, 2001; Smith, 1999). Nothing could be further from the truth (Hyten et al., 2017; Ludwig, 2018; McSween, 2003).

"Behavior" refers to "*acts* or *actions* by an individual that can be *observed by others*" (Geller, 1996, p. 115). In their training materials, one of our accredited sites (Costain Behavior Management Team) define "behaviors" as "*what we do and say*." When safety management programming focuses on reinforcing safe behavior (e.g., a driver presses the forklift horn when approaching an intersection), or any behavior that positively impacts safety programming (e.g., talking to peers about risk; inspecting equipment), substantial decreases in injuries can occur (Agnew & Daniels, 2010).

The clear link associating behavior with injuries illustrates a need for companies to focus their attention on safe behaviors. Behavior analysis

Science Moment 1.5
Environment Selects Behavior

The perspective in behavioral science is that behaviors are a product of their environment. Behaviors are initially influenced by our biology (i.e., phylogenic selection) and history of events over our lifetime (i.e., ontogenetic selection). Our early vocalizations as an infant were reinforced by our parents' praise; and now we talk all the time! We call this our learning history, which provides the foundation for how we may behave in a new situation; but the rest is based on what is going on in the moment. For example, drivers have a learning history of stopping at red lights, going at green lights, and yielding at yellow. When a driver approaches an intersection and the light abruptly changes from green to yellow, the driver's behavior responds to the environmental conditions happening at the time based on previous experience with consequences. The light turns yellow, and the intersection is still far away, with other cars ready to enter, signaling the potential for an accident. Alternatively, the driver may be running late and is close to an almost empty intersection, signaling the potential to save time. Thus, these environmental variables "select" the behavior because of the relationship between the environment and its consequences; they thus become part of the person's learning history and will likely influence their behavior in the future. Behavioral safety identifies and modifies these environmental variables.

provides the scientific foundation for programming safety management systems to engage in sustained behavior change. Behavioral systems analysis provides additional analysis and design tools for the changes needed to maintain and continuously improve behavioral programming.

This book seeks to provide this scientific foundation for basic components of behavioral safety, including risk analysis and pinpointing; direct observation and performance feedback; reinforcing process engagement; trending and functional analysis; intervention techniques; and process evaluation. We provide evidence and discuss best practices from successful behavioral safety programs that have been accredited by the CCBS Commission on Behavioral Safety. But first, let's understand the components of behavioral safety that will form the outline for the presentation of scientific principles and best-practices.

THE EVERGREEN MODEL OF BEHAVIORAL SAFETY

Behavioral safety is an empirically validated system of maintaining safe behavior in the workplace based on the work of B.F. Skinner and W. Edwards Deming (Geller, 2005a). Variations of behavioral safety have been documented to be effective in improving safe behavior (e.g., Fante et al., 2010; Grindle et al., 2000; Hermann et al., 2010; Komaki et al., 1978; Ludwig et al., 2002; Ludwig & Geller, 1997, 1999, 2000; Stephens & Ludwig, 2005; Sulzer-Azaroff & Austin, 2000)[SM-1.6].

Science Moment 1.6
Navigating Scientific Literature

Throughout this book, we cite experimental studies, field reports, books, and other materials providing evidence for the principles and processes that we describe. We encourage you to look up these citations and others we provide to further your education in behavioral safety. We provide a complete reference section at the end of the book. Some material will be readily available through an internet search engine (e.g., Google). Two additional free services we recommend are Google Scholar and ResearchGate.

Most variations of behavioral safety contain peer-to-peer observations and feedback. In this method, employees observe each other's safe and at-risk behavior while on the job and record the results on a behavioral checklist. Observers then discuss the results with their peers and provide praise for safe behavior and corrective feedback for at-risk behaviors. Behavioral checklists are then collected and analyzed to identify and intervene on the causes of at-risk behaviors. Peer-to-peer observations and feedback provide key advantages. They allow for individualized immediate feedback, which has been shown to generate the greatest amount of behavior change (Daniels & Bailey, 2014; Ludwig et al., 2010).

Several experts have outlined the basic elements of behavioral safety, which add scientific method to the peer-to-peer feedback process. Table 1.1 contains summary information of several proposed components of behavioral safety highlighting the seminal works of Sulzer-Azaroff and Austin (2000), McSween (2003), and Geller (2005a). We offer an updated methodology based on their work and the research and practice of many since.

Our Evergreen Model of Behavioral Safety is based on a review of relevant research and practice. We outline behavioral safety as a continuous process (see Figure 1.4) that includes:

- risk analysis and pinpointing;
- direct observation;
- reinforcement engagement;
- trending and functional analysis;
- behavior change interventions; and
- evaluation.

We call this model "Evergreen" because, in addition to the components of behavioral safety reviewed above, it contains iterative (repeating) "loops"

Table 1.1 Summary of Behavioral Safety Program Components

Authors	Components
Sulzer-Azaroff and Austin (2000)	• Identify behaviors that impact safety. • Define behaviors for precise reliable measurement. • Develop mechanisms for measuring behaviors to determine the current level and set goals. • Provide performance feedback. • Reinforce progress.
McSween (2003)	• A behavioral observation and feedback process. • Formal review of observation data. • Improvement goals. • Reinforcement for improvement and goal attainment.
Geller (2005a)	• Define safe target behaviors. • Observe behavior. • Intervene to increase safe behavior. • Test for behavior change.

forward and backward across the process, continuously defining, implementing, evaluating, and improving. Even when the process is a success and we've documented positive behavior change that reduces risk, we loop back up to the beginning to set new behavioral goals (pinpoints) and start anew, making the process "evergreen."

In a typical behavioral safety process, risk analysis and pinpointing are conducted first to identify potential candidates for observation and the development of behavioral pinpoints (Step 1). Next, the workforce engages in processes to directly observe behavior (Step 2). This observation of behavior allows for the delivery of feedback to differentially reinforce safe behaviors on a task while suppressing at-risk behaviors (Step 3). The behaviors of employees and managers that participate in the process also need to be driven through reinforcement (Step 4). To understand why behavioral variance trends emerge, functional analysis is conducted (Step 5). Implementation of targeted feedback, contingencies, and systemic interventions is based on the results of functional analysis (Step 6). Using the ongoing behavior observation process, behavior change is observed during ongoing time-series evaluation (Step 7). Although these steps are listed sequentially, the Evergreen Model of Behavioral Safety process is not linear.

During each step in the Evergreen Model of Behavioral Safety, decisions must be made to determine if the behavioral safety program should move to the next step. For example, if an intervention does not result in behavior change (Step 5), the intervention should first be scrutinized by measuring process data assessing the integrity of activities to determine if these occur

**Risk analysis &
pinpointing**

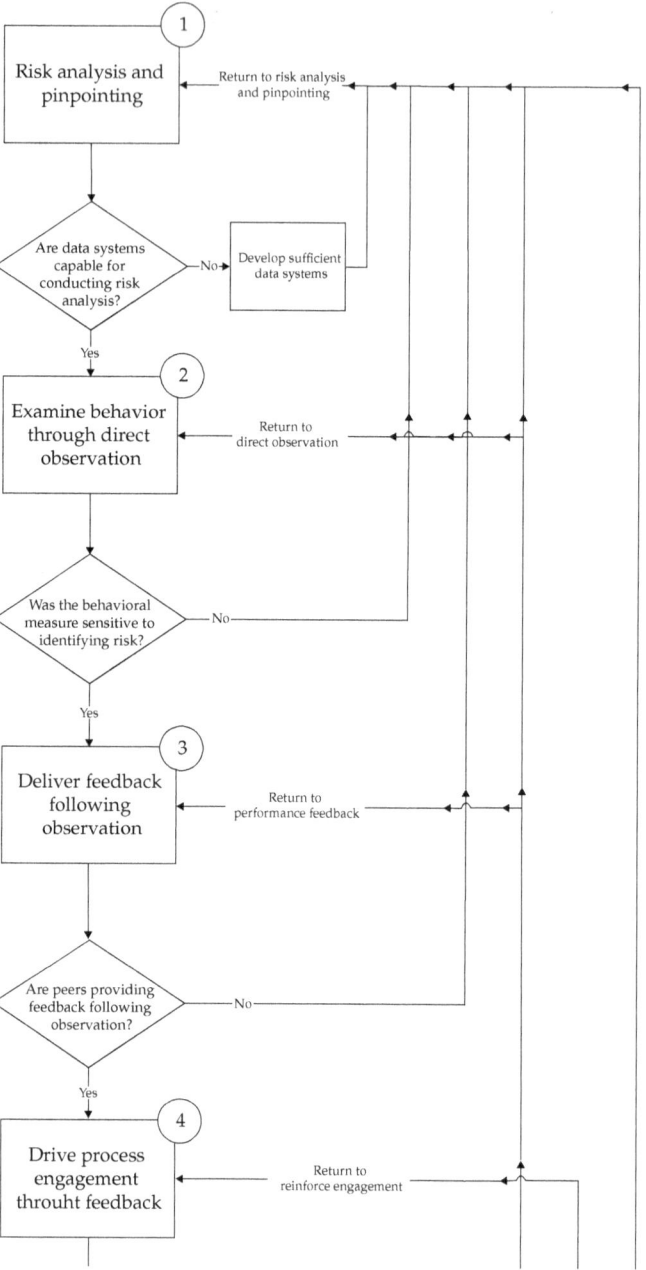

**Direct
observation**

**Performance
feedback**

**Reinforce
engagement**

Figure 1.4   Evergreen Model of Behavioral Safety

The decision points displayed between steps are not exhaustive and only serve as examples.

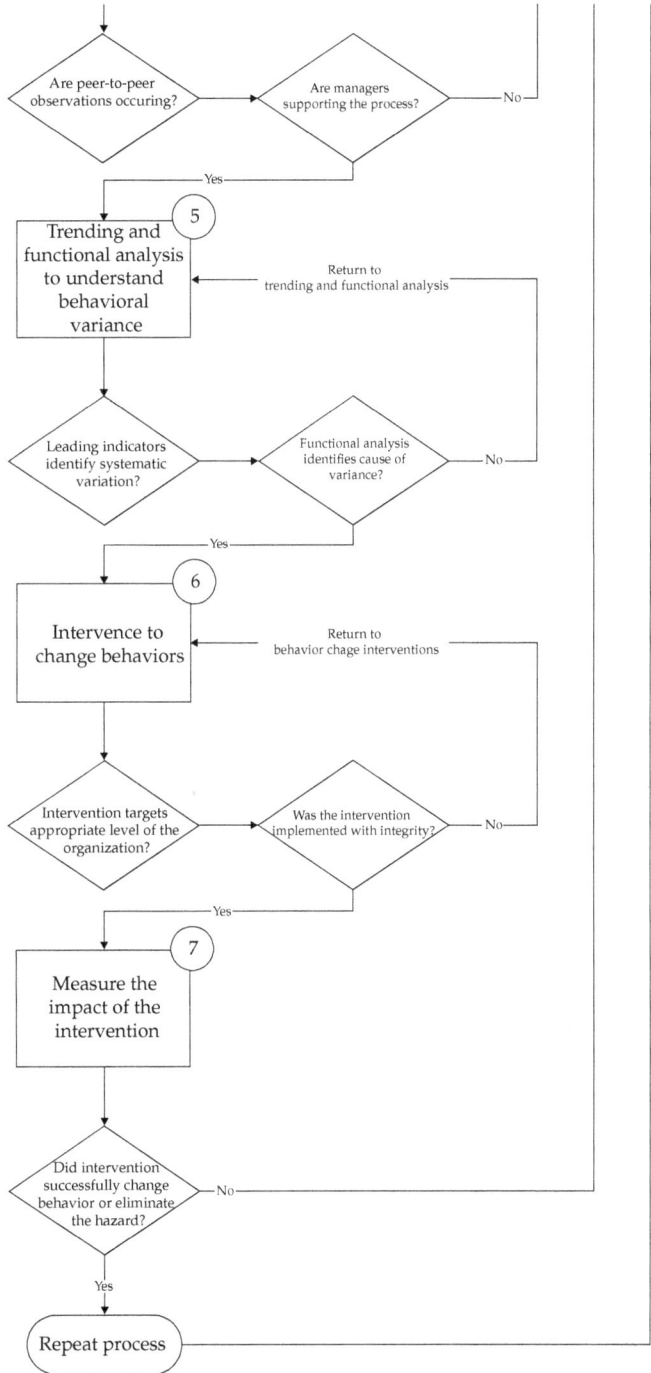

Figure 1.4 Continued

with the frequency, quality, and intensity as designed (cycling back to Step 6). As another example, if ongoing observations suggest interventions do not sufficiently change the at-risk behavior, then functional analysis (e.g., Antecedent-Behavior-Consequence analysis; behavioral systems analysis (McGee & Crowley-Koch, 2021); Step 5) must be revisited to consider additional contingencies and systems. Finally, if behavior change is demonstrated through ongoing evaluation or if the hazard has been eliminated, the process cycles back to the beginning by identifying other at-risk behaviors and the adaptive process starts again (Step 1).

The research and best practices of each of these components are discussed in the following chapters with empirical field evidence provided by the work of the CCBS (behavior.org). As noted previously, the CCBS established the Commission on Behavioral Safety in 2005 to recognize the world's best-in-practice behavioral safety programs and disseminate their methods and results. The Commission on Behavioral Safety accredits these programs using standards based on the foundations of behavior science and evaluates actual reductions in injuries among the participating workforce(s) as a function of behavioral safety programs. These standards are available on the CCBS website (behavior.org) through the "Safety" help center.

The accreditation process involves the evaluation of the applicant's behavioral safety program through interviews, document reviews, and safety metric compilations. Members of the Commission (the accreditation review team) then visit the industrial site where the behavioral safety program is active to conduct observations, data reviews, random interviews, leadership assessments, and reviews of other safety management systems; and to witness a behavioral safety team meeting. The accreditation review team then documents a program description of the behavioral safety program and writes a program review to evaluate best practices and provide recommendations for improvement. We also challenge the program to innovate further and teach us new ways of translating the science into best practices. Evidence provided in this book comes from the CCBS accreditation process; full program descriptions and reviews can be found on the CCBS website (behavior.org) in the "Safety" help center.

2 RISK ANALYSIS AND PINPOINTING

The message from the corporate safety office was that they were not so much concerned about personal safety total recordable incident rate (TRIR) metrics. They had gotten the TRIR down so low for their industry that they created their own metric to count everything—every bump and bruise, cut and stub—as an injury, just to have enough things to look at for their investigations. They even had a number of plants accredited for this performance. The word now, and rightly so, was that they were putting much more focus on reducing serious injuries and fatalities (SIFs). They certainly had their "life-critical" procedures to comply with; but the big injuries seemed to be things they hadn't considered or straight-out violations of the life criticals. They also knew they had invested big in behavioral safety and needed their teams to specifically begin observing behaviors that could prevent SIFs. But how? Where to start?

When we use behavior analysis as a basis for safety management, it requires us to determine, verify, and shape behaviors required for safe work. Therefore, a necessary and vital first step is to arrive at the critical operational definitions of behaviors for observation, analysis, and shaping. This step has earned the popular name "pinpointing" to describe the process of identifying specific, observable, and measurable behaviors which lead to desirable safety outcomes (Agnew & Daniels, 2010; Geller, 1996; McSween, 1995).

Hyten (2009) argued that desired business results should be linked back to performer behaviors. To this end, a first step in behavioral safety is to identify the response classes[SM-2.1] most related to injuries and injury reduction. This involves analyzing measures and operational considerations to identify the most effective pinpoints related to future risks in the workplace. These pinpoints are then programmed into the behavioral operations discussed later.

The process of pinpointing typically involves subject-matter experts (SMEs) who conduct ongoing risk analysis and behavioral classification. An external or internal (i.e., safety manager) expert with experience in behavioral safety will facilitate group sessions with other SMEs with more extensive experience with the jobs and tasks engaged in by the workers. While this can include management and engineers, those with the most dynamic experience tend to be workers who engage in the work.

SMEs identify and define behaviors associated with possible injuries. Particular weight should be given to activities related to potentially serious

DOI: 10.4324/9781003290711-2

Science Moment 2.1
Response Class

Behaviors start as neurological impulses that activate a muscle, which combines into moving a group of muscles, which combines into moving your body. We can describe behaviors as a single body movement, but sometimes it takes many body movements to do something. Then once we do something, we often have other things to do which combine into accomplishing something. These single accomplishments add up to bigger accomplishments. Where do we define behavior along this continuum, from micro to macro? A "response class" is a group of behaviors that produce the same effect on the environment. Behaviors in a response class serve the same function to produce a certain outcome. For example, there are many different ways to protect your back when lifting a heavy case. You could bend your knees to pick up the case; hold the case close to your body; stand up fully before twisting; and pivot your feet to turn. Each of these individual behaviors functions to protect your back while lifting. So, we have a choice to make as to where we define the response class depending on what we are trying to accomplish.

injuries and fatalities (Martin & Black, 2015). To this end, SMEs analyze past incidents and conduct other diagnostic analyses for behavioral connections in a process called "risk analysis."

RISK ANALYSIS

Risk analysis typically starts with a review of recent injuries (Wilder et al., 2018). If the company has a robust system that encourages employee reporting of minor injuries (i.e., not requiring a healthcare visit or prescription) and close calls (i.e., the release of hazard energy without contacting the body), then these more numerous leading indicators should be included in the risk analysis (McSween & Moran, 2017).

In addition to analyzing prior safety incidents, risk analysis should look at measures of operational processes that create the product or service for the organization. These organizational practices can set the environment for higher levels of risk in the workplace. Ezerins et al. (2022) proposed a safety measurement framework suggesting relationships between organizational variables and safety outcomes. Several of these variables—including production (e.g., volume of work); maintenance (e.g., repair planning, preventive maintenance schedules); labor (e.g., shift scheduling, overtime; Folkard & Lombardi, 2006); engineering (e.g., equipment and process design; Liu & Tsai, 2012); procurement (e.g., replacement parts, new equipment); projects (e.g., construction phases); and financial decisions (e.g., budget, costs)—can have a substantial and statistically significant impact on the risk levels required of frontline workers. Our behavioral safety lab at Appalachian

State University has analyzed useful measures from operational processes that can come from other safety activities (e.g., inspections, employee concerns), including maintenance metrics reflecting outstanding equipment defects (Ezerins et al., 2020); labor hours and overtime (Laske et al., 2020); engineering variance; human resources data (e.g., absenteeism, turnover, new employees, leadership change); shift scheduling (Laske et al., 2022); financials (e.g., budget and cost variance); and environmental events (e.g., wind levels; Hinson & Laske, 2020).

Descriptive and diagnostic analytics (Huang et al., 2018) are then used to trend past injuries (and minor incidents) and reveal correlations with data coming out of operational processes. This information helps us arrive at the most likely work tasks and activities related to risk in ongoing operations. For example, Mollicone et al. (2019) predicted hazardous hard-braking events for commercial truck drivers by identifying employee fatigue (e.g., through a wrist-worn device to measure sleep durations). In the construction industry, decision tree and fault analysis have been used to identify variables predictive of fall accidents (e.g., restrictive walkways, ladder hazards, weak roofing materials; Chi et al., 2014; Mistikoglu et al., 2015). Ajayi et al. (2020) modeled employee demographics and the type of task to predict injuries and accidents in the electrical power infrastructure industry.

PINPOINTS

The results of risk analysis help identify the behaviors most related to safe operations unique to your workplace, crews, and tasks. We want to pinpoint behaviors that will best provide for direct observation, revealing analysis, and effective intervention.

Group processes are typically used to pinpoint. Participants in the pinpointing should include workers who do the hazardous tasks where observations will occur (Sulzer-Azaroff et al., 1990). The participants can include employees (Cooper, 2006), managers and safety personnel (Reber & Wallin, 1984), or a combination of both (Myers et al., 2010). McSween (2003) recommends that participants in the pinpointing meeting include personnel from safety, management, engineering, and operations. The advantage of including personnel from different organizational functions is to gain their unique perspective on the contingencies in the work environment.

Employee participation in the pinpointing process is a common recommendation in behavioral safety (Depasquale & Geller, 1999; Geller, 2005b; Ludwig & Geller, 1997; McSween, 1995, 2003). An employee's perspective during the pinpointing meeting is advantageous because they have direct experience of, and interactions with, hazards. The advantage of the manager's perspective is experience in intervention design and procedure formalities (Cooper, 2006; Zohar, 2002; Zohar & Luria, 2003).

The pinpointing meeting should start by reviewing the results of the risk analysis. Then, participants are prompted to identify critical tasks and behaviors that are related to these areas of risk. After the group has brainstormed pinpoints, they can be prioritized based on:

- relation to safety (accident records);
- frequency of task occurrence;
- overlap between pinpoints; and
- feasibility of observing (McSween, 1995).

Science Moment 2.2
Operant

Skinner used the term "operant" to describe behavior as "operating" your body based on the specific situation in order to produce something. An operant focuses on body movements that have the desired effect on the environment. You may have noticed that the operant shares a similar definition with a response class that we defined earlier (see Science Moment 2.1). Responses within the operant are all body movements; while response classes can be broader to describe the results of a bunch of different operants.

Science Moment 2.3
Behavior Chain

A "behavior chain" is a set of behaviors that must occur in a sequence, all of which result in a reinforcer (reinforcers are written as "SR") when completed. In a behavior chain, the consequence of completing the behavior is the opportunity to do the next behavior, which in turn serves as a discriminative stimulus for the next. As a simple example, to drink from a water bottle on a hot day, several behaviors must occur in sequence. We have diagramed the water bottle example below:

S^D (sight of water bottle): R (pick up bottle) → SR/S^D (bottle in hand): R (take off cap) → SR/S^D (lid removed): R (drink water) → SR (water)

In this example, the consequence of having the bottle in hand reinforces picking up the bottle, but also serves as a discriminative stimulus for taking off the cap. Removing the lid reinforces taking off the cap and signals the availability of reinforcement for drinking the water.

Behavior chains can include many or few behaviors. For an assembly line employee, each step required to build a part is included within the larger behavior chain:

S^D (sight of part): R (complete step 1) → SR/S^D (step 1 complete): R (complete step 2) → SR/S^D (step 2 complete): R (complete step 3) → SR (part assembled)

The behavioral term "operant"[SM-2.2] was used by Skinner (1938) to define response classes in terms of responses with a similar common effect on the environment. Behaviors that have a common effect on the environment can be described as functionally similar (Keller & Schoenfeld, 1950). For example, bending, reaching, grabbing, pulling, turning, and releasing are all part of a behavior chain[SM-2.3] that functionally picks up and discards scrap in a work area. Similarly, the response class of "discarding scrap" could be a part of a larger response class known as "housekeeping," which includes other subordinate classes such as "replacing tools" and "removing obstacles to egress."

The goal of behavioral safety is to help workers (and managers) identify the behaviors required to do a hazardous task safely. We do this by understanding and changing the environmental contingencies that make up the context around the behavior The "context" for the behavior typically includes the instructions for the task; the physical and social environment; the workflow chain (i.e., when completing one step of a task, discriminate the availability of reinforcement for completing the next step); and other prompts designed into the setting (e.g., signs, supervision, or fellow workers). Thus, a "discriminative pinpoint" describes both the operant action and the context that discriminates that action. It tells us not only *what* to do, but also—importantly—*when* is the right time to do it.

Science Moment 2.4
Three-Term Contingency

Behaviors occur in environments which are influenced by the context provided by the setting surrounding the worker, such as equipment, tools, instructions, team members, supervision, and the like. These environmental factors, called "antecedents," serve to direct the behavior, to help the worker identify what the right behavior is in this context. For example, the operating procedures, supervisor instructions, tool setup and machine condition all serve to tell the worker, "This is a good time to do this behavior." The worker then experiences a consequence of their action: the work gets done, the machine is working again, and the boss expresses happiness with their work. When the antecedent successfully discriminates the correct behavior leading to a desirable consequence, we call these antecedents "discriminative stimuli" (refer back to Science Moment 1.4 for an overview of discriminative stimuli). For example, your supervisor tells you to use a 4-inch clamp on a hose to stop a leak. You retrieve the clamp and successfully stop the leak. The supervisor's instruction was the discriminative stimulus directing your behavior to use a specific tool, which resulted in the consequence of successfully stopping the leak. We call this whole relationship a "contingency" because the behavior is contingent on a clear discriminative stimulus and the consequence is contingent on the behavior. To understand behavior, you need to understand the contingency. We refer to the collection of those antecedents, behaviors, and consequences that make up one discriminated operant as a "three-term contingency."

In their seminal behavioral safety publications, Komaki et al. (1978) and Fellner and Sulzer-Azaroff (1984) created discriminative pinpoints targeting mostly task-specific behaviors. Examples included "When pulling dough trough away from dough mixer, hands are placed on the front rail of the dough trough and not on the side rails," and "Towmotor operators should blow their horn prior to entering the roll-wrap room or going to the dock from the roll-wrap room."

As the practice of behavioral safety progressed, less discriminated and more general pinpoints of response classes were adopted. In some cases, the pinpoint described a variety of behavioral topographies or movements. For example, the pinpoint "body positioning" related to specific instances of reaching, pulling, bending, and twisting that could put the body at risk of muscular-skeletal injury (National Institute of Occupational Safety and Health, 1997).

In other cases, the pinpoint described the outcome of behavior. Personal protective equipment (PPE) became a common pinpoint to classify an outcome of behaviors resulting in wearing PPE at critical times. Outcome pinpoints are often adopted to describe a variety of actions that engage safety features when using equipment, such as in a motor vehicle (e.g., safety belt and turn signal use; Ludwig & Geller, 2001).

Finally, pinpoints can also be written to describe environmental conditions generated as a result of behavior (e.g., "obstruction of exits" and "hazardous material storage"; Sulzer-Azaroff & De Santamaria, 1980). In most cases where outcome-based pinpoints were used, specific discriminated examples were added to training and documentation (McSween, 2003).

So—which are better:

- response class pinpoints, which group a lot of functionally related behaviors together and describe the outcome; or
- discriminative pinpoints, which typically target specific behaviors within specific tasks?

Should we try to be inclusive of many possible behaviors across tasks in a workplace using general checklists that list outcomes of behavior (e.g., body position, housekeeping, PPE); or target behavior much more related to movements and tasks (e.g., "Cut away from the body with your thumb on the back of the blade") to better discriminate the action desired? Wirth and Sigurdsson (2008) called for research to determine whether behavioral observation checklists should pinpoint task-specific (discriminative pinpoints) or general response classes of behavior. And, frankly, we are still doing that research; but here are some considerations of the pros and cons of each. Discriminative pinpoints and response class pinpoints are likely to prompt the observer to observe different behaviors. A functional pinpoint should ensure reliable discrimination of what behaviors to observe. We've argued that a discriminative pinpoint is more likely to function as a discriminative stimulus directing the observer's behavior, as the pinpoint specifies the context, specific movements, and outcomes that the observer should look for. By contrast, a response class pinpoint does not provide enough context to discriminate specific behaviors to observe. However, the benefit of the response class pinpoint is that it may prompt the observer to observe a wider variety of behaviors compared to the discriminative pinpoint, which directs observers toward one behavior. The differences in pinpoints are also likely to influence the feedback the observer provides following the observation. Finally, data obtained from these pinpoints are likely to result in different information that will influence intervention design. For a more complete discussion of discriminative and response class pinpoints, we wrote a whole other paper on the matter (Laske & Ludwig, 2022a).

If a pinpoint represents too broad of a response class, the worker's behavior (and the observer's) may not be discriminated. When this happens, we get behavioral variation, because the worker will try various behaviors to get to a personally desirable consequence (e.g., to get the job done safely—or, more likely, to get it done quicker and easier). Different workers may try different behaviors, which creates risk. For example, an outcome-based response class pinpoint such as "needle disposal" is not discriminated by antecedents in the worker's immediate environment. Therefore, the nurse may recap the needle with her other hand (at-risk); throw the exposed needle in the trash (at-risk); or discard the needle in the biohazard container (safe).

In contrast, pinpoints more specific to the task environment—such as "immediately discard catheter needle in a biohazard container after removing from patient" (Stephens & Ludwig, 2005)—are more likely to be related to proximal antecedents such as the task tools (i.e., catheter needle), equipment (i.e., biohazard container), and workflow chain (i.e., after removing from the patient) that would be more functional in discriminating the behaviors needed to perform the task safely. An excellent example of discriminative pinpointing can be found in specific descriptions of safe behaviors in cooking tasks, such as "Stand back when opening combi-ovens" used by Lebbon et al. (2012) to discriminate behaviors around potential burn incidents.

On the other hand, discriminative pinpoints may only apply to a narrow set of workers doing a narrow set of tasks. Most industrial sites have a myriad of different crafts doing a myriad of different tasks. The use of more general response classes as pinpoints in behavioral safety allows for groupings of behaviors to be considered when conducting observations across a variety of tasks and work processes. You can have one pinpoint that applies to most, if not all, types of jobs at the site. In construction, for example, "body positioning" can be applied to the earth-works folks digging holes as well as to electricians pulling conduit.

Behavioral safety programs can also use both general response classes and more discriminated operants in their processes. The most common use of pinpoints is on behavioral observation cards which serve as checklists to direct coaches engaging in direct observation and feedback with workers (described in the next chapter). The AWARE behavioral safety program at Marathon Petroleum's St. Paul Park refinery uses general response classes in its observation process. However, when risk is observed in one of the response classes, more discriminated pinpoints are adopted to identify the specific operant(s) related to the risk. For example, when "body mechanics" risk was identified in 10% to 15% of observations, their team adopted more specific pinpoints of proper extending when working with certain equipment. This greater degree of discrimination led to more targeted analysis and intervention. We will tell you this story in greater detail with data in a later chapter.

Creating a Response Class Pinpoint

Response class pinpoints should describe the product of many different behaviors that are functionally related in producing a desired outcome (e.g., wearing a safety harness). We want to avoid "lagging indicator" organizational results (e.g., injury reduction) or simply describing changes in a performance metric (e.g., safety inspection results, number of behavioral safety observations). Laske & Ludwig (2022a) offer the following criteria for response class pinpointing for use in behavioral safety observations, coaching, and analysis:

- Valuable contribution to organizational result (e.g., related to injuries, especially SIFs): We must ensure that the sum of the behaviors in the pinpoint actually produces something valuable. In behavioral safety, this means our risk assessment and pinpoints should identify behaviors critical to safe operation and the avoidance of SIFs. In response class pinpoints, we must therefore identify the behavior categories that drive down injuries.

- Work output: Carl Binder, a former student of B.F. Skinner, has conducted decades of work with large companies trying to improve performance over a wide array of key performance indicators beyond safety. Based on his experience, he suggests we pinpoint using the "output of behavior"[SM-2.5] to best describe the response class. How do you determine the difference between a behavior and output of behavior? Thomas Gilbert, an influential behavioral engineer, has offered the following quote that we find beneficial: "Behavior you take with you, accomplishment you leave behind!" (Lindsley, 1991, p. 458). Thus, instead of listing the actual behaviors in a response class pinpoint, we should list the output that all those behaviors produce.

Science Moment 2.5
Output of Behavior

Experts in behavioral systems analysis such as Maria Malott (2003) and Dale Brethower (1972, 1982) outline the basic building blocks of a process as follows:

Input – throughput (process) – output

and relate them to behaviors. They recognize that behaviors are little processes in and of themselves. Behaviors have inputs (e.g., environmental antecedents, cultural antecedents) and outputs (e.g., personal, group and organizational consequences). Individual behaviors have outputs; as do response classes, which are made up of a bunch of functionally related behaviors all operating toward the same goal: the output of behavior.

Based on this advice, we should avoid using examples of behavior (e.g., writing comments on the observation checklist during behavioral observations) or using passive verbs to describe the pinpoint (e.g., "Observation completed"). What we are not doing here is just using the safety rules as we make our pinpoints. Also, remember: no adjectives! If pinpoints use adjectives, such as "appropriate" or "excellent," the observer will then be required to do a subjective evaluation within the observation, hurting our reliability. We should instead pinpoint the valuable product of that behavior (e.g., identifying safety risks).

- Process (action verb): After specifying the work output, the process that produces that output must be defined. The process can often be defined as an entire response class of behavior. Defining the process as an action allows for observation, coaching, and behavior analysis to occur.

- Reliably measured: Any pinpoint must be able to be reliably measured (DiGennaro Reed et al., 2018). To this end, the pinpoint should be objective and sufficiently defined so different observers will report seeing the same thing. This is known as "intra-observer reliability" in research and is a big deal because it proves that our observations are not biased by an individual observer. We also want to make sure that the observations of the pinpoint don't change across time, as it might drift with familiarity, fatigue, or change based on the situation. Avoid using adjectives because they require the observer to make a subjective judgment. For example, "effective [adjective] communication with dispatchers" leaves the observer to subjectively define "effective." Requiring subjective judgment will diminish reliability. Finally, Binder (and our science) insists that pinpoints must be able to be counted. Therefore, the outputs of behavior must be described as countable units.

- Under the individual's control: One challenge we have with response class pinpoints is identifying what level of specificity to use. The pinpoint can describe anything from a task outcome (e.g., "Machine is locked out and tagged") to the performance of the process (e.g., the time it takes to complete the maintenance task on the machine), all the way up to organizational levels (e.g., "Machines are operating reliably"). Which do we choose? If we go high level, individuals will have a hard time seeing how they can have meaningful control over the outcome. If we go too low, we may not have a pinpoint that is generalizable across the plant. Thus, Austin (2000) suggests listing all levels while pinpointing and then deciding which behavioral output is still under the control of the worker yet valuable to the organization.

Based on Binder's (2016; 2017; 2022) recommendations, here is a rubric for writing response class pinpoints:

- Action verb – the process:
    "Lock and tag during maintenance work orders."
- Work output – output unit as a noun that can be counted:
    "To disengage machine energy on 100% of maintenance work orders."
- Valuable contribution – how the outcome adds value to an organizational result:
    "To reduce the risk of a serious injury or fatality."

Creating a Discriminative Pinpoint

One of the most influential texts in behavior analysis was authored by James Johnston and Hank Pennypacker (1980), who wrote the esoteric statement:

> The behavior of an organism is that portion of an organism's interaction with the environment that is characterized by detectable displacement in space through time of some part of the organism and that results in measurable change in at least one aspect of the environment (p. 48).

This is a fancy way of saying that a person moves their body and, as a result, changes something in the environment around them.

We adapted Johnston and Pennypacker's (1980) definition of "behavior" to propose six criteria for a discriminative pinpoint (Laske & Ludwig, 2022a). We also cite a bunch of other researchers who agree with us. The first four describe what should be in the behavioral description and the final two you'll recognize as summarizing the overall criteria for all pinpoints regardless of specificity. A discriminative pinpoint should contain the following elements:

- Bodily (or verbal) action (Johnston & Pennypacker, 1980; Mayer et al., 2019; Miller, 2006; White, 1971): A simple way of assessing if your pinpoint is indeed an action is using what we call the "dead-person test"[SM-2.6] (Lindsley, 1991; Ludwig, 2018). We should pinpoint the active behavior required to be safe. That's right: list the safe behavior instead of the at-risk behavior variant. This way, we are always teaching the desirable behavior with a kind of positive vibe. Listing pinpoints as the safe behavior assures that every time we use the pinpoint in our safety management systems (e.g., training, prompts, checklists, feedback), we are discriminating the specific safe behavior desired. Why would we want to discriminate the at-risk behavior that is not desired? Pinpointing active safe behaviors helps communicate with employees what a person *should do* instead of telling them *what not to do* (McSween, 1995; 2003).

Science Moment 2.6
Dead-Person Test

The dead-person test was taught by Ogden Lindsley (a B.F. Skinner student at Harvard) in the 1960s to help teachers come up with observable pinpoints to reliably observe their students' behavior. Here is the test: *if a dead person can do the pinpoint, then it is NOT a behavior . . . because they are dead.* Dead people cannot behave. This helps us avoid pinpointing the ommission of something (e.g., "Failed to comply with . . ."). It also helps us avoid pinpointing an output (e.g., wearing PPE – I'm sure some dead person somewhere is wearing PPE in a coffin). In addition to "wearing," other desired conditions like "Orderly storage of tools," "Correct body position," and "Completed permit" are all outcomes of behavior. They may be available to direct observation—you can see the harness on the individual—but you are not observing behavior. Rules and compliance-based pinpoints often fail the dead-person test. You may pinpoint "Failure to set the equipment guard" as a violation of a rule. But it is impossible to see the worker "not put up the guard." Instead, you can only note the omission of the desired behavior of "lifting the guard."

- Behavior environment interaction (Johnston & Pennypacker, 1980): Bodily movement changes the environment around it. What did the behavior move or manipulate? A tool? A part? A pencil?
- When the behavior should occur (Kazdin, 1994; Mayer et al., 2019; McSween & Moran, 2017): This is an important one, which is often overlooked. When we describe the *when*, we are describing the environmental

conditions that should discriminate the behavior. That's right: we are listing the discriminative stimulus as the antecedent that says, "Now is the best time to do this behavior to be safe."

- What the behavior will accomplish (Mayer et al., 2019; Miller, 2006): Now we get to relate the specific behavioral pinpoint to Binder's functional output of this behavior and others in its response class. This also describes the consequence of the behavior.
- Reliably measurable (Chance, 2006; Johnston & Pennypacker, 1980; Miller, 2006; Sulzer-Azaroff & Austin, 2000; Sulzer-Azaroff & Fellner, 1984): The pinpoint must also be measurable—duh!
- Under the employee's control: In addition, the pinpoint should be under the individual's control (Daniels & Bailey, 2014; Sulzer-Azaroff & Fellner, 1984).

These elements of a discriminative pinpoint describe the three-term contingency central to behavior analysis (Skinner, 1953; also see Science Moment 2.4). Yup, we just did the Antecedent-Behavior-Consequence analysis in our pinpoint! The bodily action is the behavior operating on (this is why we call behavior the "operant") something physical in the environment (e.g., tool, product, equipment). This environmental object also serves to discriminate the specific tool, piece of equipment or whatever the worker is to use to best stay safe. Then we drop in a temporal cue by saying when the behavior should occur within the workflow to further discriminate the exact time the behavior will keep you safe. Finally, the pinpoint explicitly states the desired consequence accomplished by the behavior. If the worker does the behavior but doesn't accomplish the desired consequence, well, we need to change the pinpoint to help the worker succeed!

With all this in mind, consider the following sentence structure for discriminative pinpoints Ludwig (2018):

- Do what (action verb):
  "Cut, pushing away from the body with thumb behind the blade."
- To what (subject of the action – physical thing in the environment):
  "The tape securing the motor."
- When (salient discriminative stimulus—e.g., workflow step):
  "After the motor has been locked and tagged out."
- For what purpose (accomplishment consequence of the behavior):
  "In order to lift the motor out of its casing for repair."

Other examples of discriminative pinpoints (Laske & Ludwig, 2022a) using this sentence structure might include "Loosen back bolts of flange when breaking the seal to avoid potential chemical release coming toward the body" and "Place venting hose over valve opening prior to venting pressure to avoid chemical aerosol exposure."

Finding Risk

There is one final criterion that is critical for any pinpoint. The key to a successful behavioral safety program is its ability to identify the risks occurring in the workplace. If we pinpoint behaviors that seem important (e.g., PPE), but these are already being done close to 100% of the time, then why target them for behavior change improvements? Workers have already mastered the behavior. Focus on behaviors that need changing! A good pinpoint is one that, once we start observations, will help us identify the risk already out there in the workforce's behavior and provide a baseline on which to improve.

We recently conducted a study (Laske & Ludwig, 2022b) that found that pinpoints generated by managers were less likely to find at-risk behaviors occurring in the workplace. Managers tended to pinpoint behaviors around housekeeping and policies which often resulted in near-100% safe recordings of behavior when used in observation processes. In contrast, employees were more likely to pinpoint specific body movements that occurred in their immediate working environment. Across multiple departments, employee-driven pinpoints were also more likely to meet our discriminative pinpoint criteria than manager pinpoints (Laske & Ludwig, 2022b). Most importantly, when employee-developed pinpoints were used in the observation process, more of the at-risk behaviors present in the work were recorded.

When observations are done by many employees, rather than only safety professionals/managers, there are greater rates of injury reduction (Predictive Solutions, 2012). Conversely, if only safety managers conduct observations, the probability of injury occurrence is actually greater! These results support previous findings (Cooper, 2006; Ludwig & Geller, 1997) that it is advantageous to include employees in the pinpointing process.

How the pinpoint is worded is also important. Our behavioral safety lab at Appalachian State University looked at three years of behavioral observation data from multiple departments doing behavioral safety observations at a large business division of a chemical manufacturing company. In one study, we looked to see if the quality of pinpoints might be related to a reduction in the probability of an injury (Matthews, 2022). We found that pinpoints that described safe behavior instead of at-risk behavior were likely to reduce the probability of an injury when used in observations.

Yes, pinpointing is serious work and needs practice to do well. However, to truly understand the behaviors we pinpoint, they must be measured frequently and accurately (Agnew & Daniels, 2010; Geller, 1996, 2005b; McSween, 1995). Behavior analysis teaches us that direct observation of behavior is critical to understand the true reason for the behavior (Johnston & Pennypacker, 1980).

3 DIRECT OBSERVATION

Reggie had never done an observation, but he decided to give it a try after a Marathon Employee Safety Awareness member assured him that the observation card he could turn in was "no name, no blame." He chose to watch a maintenance worker doing a repair near his unit because this was not a job he was familiar with and he worried about the risks some maintenance workers encountered while doing these tasks. He told his supervisor he was going to do an observation and went up to the maintenance worker, Tracy, and introduced himself. He then asked if it was okay to do an observation and reminded Tracy that this was all "no name, no blame." He showed her the card he would be using and she agreed. He found this conversation was not as awkward as he had feared. Tracy seemed to be a bit more cautious with her work and Reggie figured this was because she was being watched. But he thought, "At least she gets a chance to try to do everything safely." He found the observation to be quite fascinating, as he started to understand why Tracy was doing what she was doing. He was a bit surprised to find that he still had a couple of at-risk behaviors listed that he could talk to Tracy about after the observation. He wrote a comment about how Tracy had to stand awkwardly to reach behind some piping to get to a valve, to provide a bit more information for his "at-risk" check next to body positioning. This led him to consider his own behavior in these situations and others he encountered. "Wow, that was the best safety training I've ever had," he thought to himself as he returned to work and found himself doing his tasks as cautiously as Tracy.

A fundamental process we insist on in behavior analysis is the measurement of behavior. Without direct observation, we are left to assume why an error occurred. I'm sure you know what happens when we ASS-U-ME. Unfortunately, our natural human biases make us more likely to attribute the at-risk behaviors we see to the fault of the person; and we tend to use labels of personal characteristics to try to explain errors (e.g., "He must be stupid"; Ludwig, 2018). But personal characteristics such as personality (e.g., sensation seeker), attitudes (e.g., renegade), or attributions (e.g., complacency) cannot be used to solve safety problems. Instead, attempts to "change the person" are antecedent based (e.g., training), and do not attempt to change the environment which *selected* the behavior in the first place (see Science Moment 1.5). Ultimately, attempts to change the person are ineffective in sustaining behavior change. You can't change a person, but you can change behavior.

DOI: 10.4324/9781003290711-3

Therefore, an important maxim of behavioral safety—at least in programs based on behavior analysis—is "no name, no blame" (Cooper, 2009; Geller, 2005b; Ludwig, 2018; McSween, 1995, 2003). When direct observation of a worker's behavior takes place, their name is typically not recorded, even if at-risk behavior is noted. Behavior is a product of its environment (Skinner, 1974);[1] thus, we would expect the at-risk behavior to be done by other workers finding themselves in the same environment. We would expect the specific at-risk behavior to be a sample of behaviors engaged in by others in the same environment. Therefore, the name of the person is inconsequential: it is not about them; it is about the behavior and the environment it takes place in. Further, recording someone's name has cultural implications because observers might fear social punishment[SM-3.1] if they "called out" a peer. If names are written down and tracked, participation in direct observation drops big time and the data we get from observations contaminated because very few people are comfortable documenting another worker's at-risk behaviors (Ludwig, 2014).

Science Moment 3.1
Punishment

Punishment is when a consequence decreases the future probability of that response. These consequences tend to be aversive. For example, an employee runs downstairs (behavior) and then takes a painful fall (consequence). If this consequence decreases the future probability of running downstairs, this is an example of punishment. Discipline is often threatened by managers when employee behavior breaks rules. If discipline is applied and reduces undesirable behavior, then the discipline is punishing. However, if employees never come into contact with discipline (either personally or through someone else), the behavior will persist. Punishment only works when it is used. However, Skinner warned against the negative side effects of using punishment to change behavior because it can result in *adjunctive behavior* where the employee acts out to punish back (e.g., spreading rumors about the boss).

Most everything you need to know about why a behavior occurs is available through direct observation (Ludwig, 2018). Behaviors are actions that you can directly experience as an observer. An effective pinpoint directs the observer to pay attention to a specific set of actions. Observers are typically provided a critical behavior checklist containing pinpoints on which they will record whether the worker performed behaviors as "safe" or "at-risk" (Geller, 1996, 2005b). The checklist also allots space for further comments to elaborate on the behavior observed, and to describe the antecedents and consequences in the environment that may be associated with the behavior (Geller, 1996, 2005b; McSween, 1995).

Indeed, a second goal of the observation is to describe the immediate environment that related to the at-risk behavior (Ludwig, 2018). When behavior is directly observed, you can note a number of antecedents in the environment such as:

- various prompts (e.g., signs, work instructions, permits);
- availability and usability of the proper tools for the task;
- the behavior of nearby peers and supervision (i.e., offers to help, social distraction);
- conditions of equipment;
- the workflow of task steps preceding;
- upset production (e.g., machine failures, backlogs, material flaws);
- apparent hazards; and
- peers taking risks (i.e., modeling).

The observer can also see the direct consequences of the behavior. These are outcomes that happen because of the behavior, *contingent* on the behavior, that may serve to reinforce or punish future behavior. For example, an observer might observe a worker jump over a trench instead of walking around it. That observer could then see that this route was faster because it was closer to the tool the worker had left behind. Therefore, the observer could understand the worker jumped the trench because it was quicker. Other consequences that could be observed might be completion of a step in the task; physical discomfort; body excursion; verbalizations by peers or supervisors (e.g., praise, socializing, scolding); access to reinforcers (e.g., break time); discipline; and, of course, injury.

WHO IS BEST TO CONDUCT THE OBSERVATION?

Really, the only criterion in doing an observation is that you have eyes and access to the work. While we can all see what is happening, various types of workers will be looking for different things when doing an observation. Engineers, for example, may be looking at the interaction of worker behavior with their engineering designs. They will be trying to determine if the engineering that went into the machines, tools, and workflow discriminates the right action on the part of the workers; or if the safe action punishes awkward or exhausting body movement. Procurement officers may wish to observe workers using tools to see if the workers can perform the desired ergonomic action to inform future purchase decisions. Managers wishing to improve work processes may want to directly observe workers within a workflow to see if they make sense and are not inadvertently causing workarounds. This practice of managers and professionals getting out of their offices and onto the front line is called *Gemba* in Japanese quality improvement literature (Imai, 1986; Mazur, 2003). It is the basis of modern process improvement methodologies (e.g., LEAN, Six Sigma), and often leads to innovative solutions (Fante et al., 2010) because these people can observe behavior in the context of critical processes.

Based on their systematic review of the literature, Boyce and Geller (2001) advocated that the best folks to do observations are the workers on the front line. Workers who do direct observations of their peers performing work tasks are in a unique position to find behavioral risks as they happen, due to their immediacy within the working environment. Managers are not. Workers are less likely to alter their behavior when being observed by a fellow employee as opposed to a manager. A manager's presence can be a discriminative stimulus that sets off a whole host of compliance behaviors in workers who wish to avoid disciplinary consequences. Finally, employees are in the best position to mitigate risk when they find it. They are more likely to understand the contingencies associated with the task and can act quickly to intervene on the spot.

The cool thing is that the peer-to-peer observation process changes the observers' behavior as well! Alvero and Austin (2004) and Alvero et al. (2008) showed that observers who watched and recorded lifting behaviors in others increased their own proper knee bending and back posture when they lifted. The same observer effect[SM-3.2] was documented when observers who recorded ergonomic sitting behaviors ended up sitting more ergonomically themselves. Ludwig and Geller (1999) asked pizza delivery drivers to record license plate numbers of other drivers who were wearing their safety belts during a safety belt campaign in which a local radio station gave out prizes to drivers whose license plate number was read out over the radio. The drivers who were doing observations increased their own belt use by 32%.

Science Moment 3.2
Observer Effect

We expect the person being observed to change their behavior due to the "observation effect," also known as the "Hawthorn Effect." Just knowing someone is watching will make you change your behavior. However, doing observation also affects the observer! The observer effect is when an observer's behavior changes after conducting an observation on another person. The observer effect is a well-documented behavioral phenomenon (Alvero et al., 2008; Alvero & Austin, 2004; Blackman et al., 2021; King et al., 2018; Sasson & Austin, 2005), suggesting that the observer is three times more likely to change their behavior than the person observed. Blackman et al. (2021) found in a series of studies that, to maximize observer effects, you should make sure your observers are well trained and accurately observing behavior. In Chapter 2, we provided the criteria for a discriminative pinpoint that may better direct observers to which behaviors to observe.

Thus, workers who engage in behavioral observations may indeed be more likely to engage in their own positive behavior change—at times even more so than the peers they observe. When workers engage in formal observations, they become more fluent in identifying risks in their own behavior as well as the behaviors of others working around them. By engaging in a formal observation, the employee has an opportunity to step away from attending to their own work tasks and more objectively observe at-risk behaviors in the context of the environment. They can see how the environment promotes the at-risk behavior (e.g., lack of proper tools). They are also in a much better position to see the potential injurious effects of at-risk behavior as other workers interact with hazards in the environment. Seeing how another worker can get injured may help the observer better understand how the hazards they encounter when doing their job may lead to an injury unless they adapt their own behaviors. Therefore, the impact of direction observation goes beyond the current context when it is conducted. Results from our behavioral safety lab revealed that when an observation was conducted, the probability of incident decreased significantly over the following three days (Ludwig et al., 2023). Specifically, across two different divisions, incident likelihood decreased 23% and 17% over that subsequent period.

The scientific explanation for this is a verbal process of rule-governed behavior[SM-3.3] (R.W. Malott, 1993; R.W. Malott et al., 1993; Skinner, 1966, 1969; Vaughan, 1989). When a worker does a task and must decide whether to engage in safe or at-risk behavior, they could sizeup the situation using the "Antecedent-Behavior-Consequence" logic. In essence, they reason, "In this situation [Antecedent], if I act this way [Behavior], then this will happen to me [Consequence]." This contingency-specifying statement is a *rule* which mediates their course of action. When we observe others, we have the opportunity to consider the rule-governed behavior that determines their actions. These *rules* then inform our personal *rules* as we go back to our work, thereby changing our behavior.

For example, when we see a construction worker jump the trench instead of walking around it, we may consider the rule "When I'm in a hurry and have left my wrench by the toolbox, if I jump the trench, I'll save time." But after standing back and watching the worker actually do this, the observer may see that the time saved is not that substantial and the potential for serious injury is more probable than previously considered. The observer may establish new rules for themselves when they engage in similar tasks in the future.

When the observer sees an at-risk behavior and considers the environmental context related to the behavior, they describe the *rule* to themselves and should write the rule that seemed to govern the behavior as a comment. However, probably the most impactful part of direct observation is when the observer discusses what they saw and the potential rule-governed behavior by giving immediate feedback to the worker (Step 2; Figure 1.4 Evergreen Model).

Science Moment 3.3
Rule-Governed Behavior

You talk to yourself, right? When you do this, you may chat with yourself about what might happen if you did something. This is how rule-governed behavior works. Rule-governed behavior is behavior shaped by verbal descriptions of contingencies (i.e., rules) in contingency-specifying statements. Consider an employee who describes a contingency to a peer about their work environment: "When the supervisor comes around [Antecedent], put your hearing protection in [Behavior], and the supervisor won't yell at you [Consequence]." If this rule is effective, it alters the likelihood of the employee putting in their hearing protection in the presence of the supervisor. In this case, the rule helps create a discriminative stimulus (i.e., the supervisor's presence) to get the employee to put in hearing protection without having to get yelled at as a consequence. This way, we can learn behaviors without having to suffer the consequences! These verbal antecedents do not always have to come from another person and can be self-generated. We do this all the time in our decision making and it may be part of what makes us human. Knowing that we all use the three-term contingency in our self-talk helps us consider why we eventually end up doing what we do.

Note

1. We find it interesting that critics of behavior analysis, more broadly, charge the field with environmentalism (Mahoney, 1989; see Todd & Morris, 1992 for a review)—incorrectly, of course (Catania, 1991; Skinner, 1974)—while behavioral safety is misrepresented as not considering environmental influences (DeJoy, 2005; Howe, 2001), albeit for different reasons.

4 PERFORMANCE **FEEDBACK**

As cool as she was looking, Tracy was apprehensive about the feedback that Reggie was going to give her after the observation. But she was curious as well. What had he seen? Did she do okay? She braced as Reggie walked back up to her with his card. She was a bit surprised when he asked her about the task she was doing. It relaxed her to talk about what she was doing and how she was doing it. He then asked about the hazards she encountered, which she rattled off quite quickly. Reggie added a couple that she hadn't considered, then asked what she had done to stay safe around these hazards. Tracy mentioned her special personal protective equipment and staying out of the line of fire in case the valve released. Then, to her surprise, Reggie listed a couple more things she had done safely: she had called the operator to double check the pressure and adjusted the angle of her tool so she would be less likely to bang her hands if she slipped. Tracy realized that she did consider her personal safety to be important and became determined to keep up these practices. As Reggie was finishing up, she was reviewing the task she had just completed and recalled that she had lost her balance slightly when reaching between the piping for the valve. Reggie said he had noticed that and asked if he could also describe some other behaviors that may have put her at risk. They then began to discuss what to write in the comments as a suggestion about what to do about the valve, so the next person wouldn't have to take the same risk. In the end, Tracy realized that she had done most of the talking. When Reggie said, "Thank you," then shook her hand, she decided it had been a rather pleasant experience and began to consider perhaps doing an observation herself in the near future.

Feedback as a construct and intervention features prominently in the scientific literature relating to industrial/organizational psychology (Luthans & Peterson, 2003; Luthans & Stajkovic, 1999; Waldersee & Luthans, 1994) and organizational behavior management (Alvero et al., 2001; Balcazar et al., 1985; Slieman et al., 2020). Giving feedback accomplishes certain critical things behaviorally.

First, feedback reinforces safe behaviors. Komaki et al. (1978) attribute the effectiveness of feedback to "positive reinforcement," which refers to a stimulus provided after a behavior that causes that behavior to increase. Geller (2005b) argues feedback can serve as social praise and thus function as a reinforcer. Reinforcement[SM-4.1] is a key principle in behavior analysis and is the key to changing behavior. Feedback noting safe behavior positively reinforces[SM-4.2]

DOI: 10.4324/9781003290711-4

when it strengthens safe behavior, making it more likely to happen in the future. Researchers frequently report robust effects of feedback modifying and reinforcing safety-related behaviors (e.g., Fox & Sulzer-Azaroff, 1989; Komaki et al., 1978; Ludwig et al., 2010; Ludwig & Geller, 2000; Rhoton, 1980). Merely praising someone for safe behavior will not act as a reinforcer if it does not make the behavior more likely to happen in the future—that's just praise. We don't care if it makes someone feel good; that's not the point. Instead, feedback should offer verbal rules that suggest that the safe behavior, done at the right time, will provide benefits to the worker. And one benefit, indeed, is getting this feedback.

Science Moment 4.1
Reinforcement

Reinforcement is when a consequence increases the future probability of a behavior (we note this as "SR"). For example, for someone dehydrated (S^D), purchasing a soda from a vending machine and drinking the soda (R) will result in them no longer being dehydrated (SR). Because of this consequence, the future probability of purchasing a soda from the vending machine when thirsty will increase. The behavior is reinforced.

Science Moment 4.2
Positive/Negative Reinforcement and Punishment

We have already defined "reinforcement" and "punishment" (see Science Moments 4.1 and 3.1 respectively). However, these can be further defined based on whether the behavior causes something to happen (positive) or causes something to go away or to be avoided (negative). In the case of positive reinforcement, the behavior results in something being added, typically pleasant. For example, an employee observes a peer cleaning up tools in the work area, says "Thanks" and buys him a coffee. As a result, the peer is more likely to clean up in the future. In the case of positive punishment, the behavior causes something aversive to happen. For example, an employee does an observation and starts giving feedback to a peer, which causes the peer to get angry and tell the employee to "kick rocks." The employee never does another observation. We call a consequence a negative reinforcer when the behavior gets rid of or avoids something, typically aversive. For example, a manager wants her employees to do behavioral observations. She makes it a point to go around daily and hound them do the observations. As a result, employees are more likely to do observations to stop the hounding. In this case, the manager's hounding is not a punisher, because the behavior increases. Finally, negative punishment causes you to lose something desirable. When a manager enjoys joking around with his crew, he is unlikely to discipline any of them for fear of losing this comradery.

Feedback can punish at-risk behavior. Now, a "punisher" doesn't involve beating someone up by scolding them until they feel bad. When you do this, they are more likely to avoid you then decrease the at-risk behavior! Instead, a "punisher" is anything that decreases behavior, which is what we want to do with at-risk behavior. The goal of feedback here is to offer a verbal rule describing how, given the hazards present, the behavior puts the worker at risk of a specific injury. If the worker is unaware of the risk, this should establish a new rule for them reducing the behavior. However, in many cases folks are taking a calculated risk in doing the at-risk behavior on purpose, because by definition, this behavior is being reinforced by something else— typically, because the behavior is easier, quicker, less cumbersome or more effective in getting the task done. The calculation is that the probability of reinforcement by doing the at-risk behavior is higher than the probability of negative consequences like injury. The feedback must counter this calculated risk by offering a new calculation. This new calculation, through the verbal rule discussed in feedback, should emphasize the likelihood of injury by pointing it out from the observer's perspective (e.g., how close their arm was to the ignition source).

Feedback on at-risk behavior should also be a negative reinforcer. Now, don't get wrapped around the axel here: "negative" simply means "to take away." When a worker engages in at-risk behavior, the feedback should point out the behavior that put them at risk and then offer the safe alternative. We want to reinforce the safe alternative, not the at-risk behavior. Therefore, the feedback on at-risk behavior discussing the safe alternative offers an antecedent directing the safe behavior in the future. The next time the observer has a chance to watch the worker do the same task (and they should endeavor to do so), they can (and must!) swoop in and offer positive reinforcement to strengthen the new behavior. However, if the worker is still engaging in the at-risk behavior, they will receive the same critical feedback again. Keep this up and the safe behavior may start happening to "take away" all the nagging!

Let's be clear: the power of feedback in peer-to-peer conversations is that it can be delivered on the spot, right after the behaviors happened. Immediate feedback is far more powerful than delayed feedback. Delayed feedback should really be called "feedforward" (Hickman & Geller, 2003), as it is more of an antecedent for the next time you do the task than something that reinforces or punishes your behavior from the last task (Agnew & Daniels, 2010; Mayer et al., 2019). Unless they are walking around, supervisors and managers typically don't learn about an at-risk behavior until after it happens: You get called into the office and receive the feedback at some point after you have done the task. In contrast, peers are working right beside you. During a formal observation, the feedback is delivered immediately. Once workers become fluent in talking to each other about clear and present risks, these informal feedback conversations can happen all the time, on the spot.

Agnew and Daniels (2010) suggested feedback should describe specific behavior the employee can directly control. Feedback reinforcing value is strengthened when paired with discriminative stimuli in the form of goals (Eikenhout & Austin, 2004; Ludwig & Geller, 1991, 1997), standards (Goomas & Ludwig, 2007, 2009), and task clarification (Merritt et al., 2019). Ultimately, feedback can be offered in a plethora of interventions, such as group feedback in comparison to established goals and graphic depictions. Group feedback on injury rates may not be effective in influencing employees to change their behavior (Komaki et al., 1978). The Occupational Safety & Health Administration (OSHA) highly discourages incentives based on injury reduction goals (e.g., number of days without an incident), because these incentives negatively punish injury reporting (OSHA, 2012, 2016, 2018). When group feedback shows the interval to the no-injury goal closing, anyone reporting an injury causes the loss of the often-valuable incentive for themselves and their peers. Punished reporting not only puts the health and finances of the injured in jeopardy, but also truncates valuable injury information that the company could use to mitigate future injuries.

Instead, offering performance feedback by focusing on pinpointed behaviors further discriminates these key safety behaviors and can provide reinforcement when the behaviors are achieved (Agnew & Daniels, 2010; Geller, 1996; Geller, 2005b; McSween, 1995). A great example comes from one of the first published studies of behavioral safety. Komaki et al. (1978) pinpointed safe behaviors in a food manufacturing plant. Behaviors such as cutting technique (e.g., "Cut with one hand, hold object above the cut with the other hand") and the "use of two people when moving conveyors" were pinpointed. Managers collaborated with employees to set a 90% goal for each behavior. Direct observation of pinpoints and frequent group feedback (three to four times a week) stating the percentage of safe behaviors were posted for the workforce. Safe behaviors increased from 70% to 96% and from 78% to 99% in two departments; but both dropped back to previous levels during a return to baseline conditions[SM-4.3]. Similar examples showing the effectiveness of group feedback on pinpointed behaviors have been reported with pizza delivery drivers (Ludwig & Geller, 1991, 1997), warehouse workers (Bateman & Ludwig, 2003; Goomas et al., 2011), anesthesia nurses (Stephens & Ludwig, 2005), retail associates (Doll et al., 2007; Eikenhout & Austin, 2004), manufacturing workers (Berglund & Ludwig, 2009), oil workers (CCBS, 2022b), coal workers (Bumstead & Boyce, 2005), roofers (Austin et al., 1996), and sports teams (DePaolo et al., 2019).

Group feedback, while effective, may only be effective on the behavior of some people, and not all. More personal, individual feedback may be

Science Moment 4.3
How to Analyze and Detect Change

To analyze and detect change, data must be studied over time instead of just looking at a single data point. Consider the example of a company trying to decrease the number of product errors in its manufacturing process. To this end, the organization implements a LEAN/Six Sigma intervention. To determine whether the intervention has been successful in creating change, the organization must look at performance before the intervention (the baseline) versus after the intervention (we put vertical lines in graphs to show when the intervention started).

Figure 4.1 contains several examples and nonexamples of detecting change. In Graph A, the change is easy to identify because of the immediate

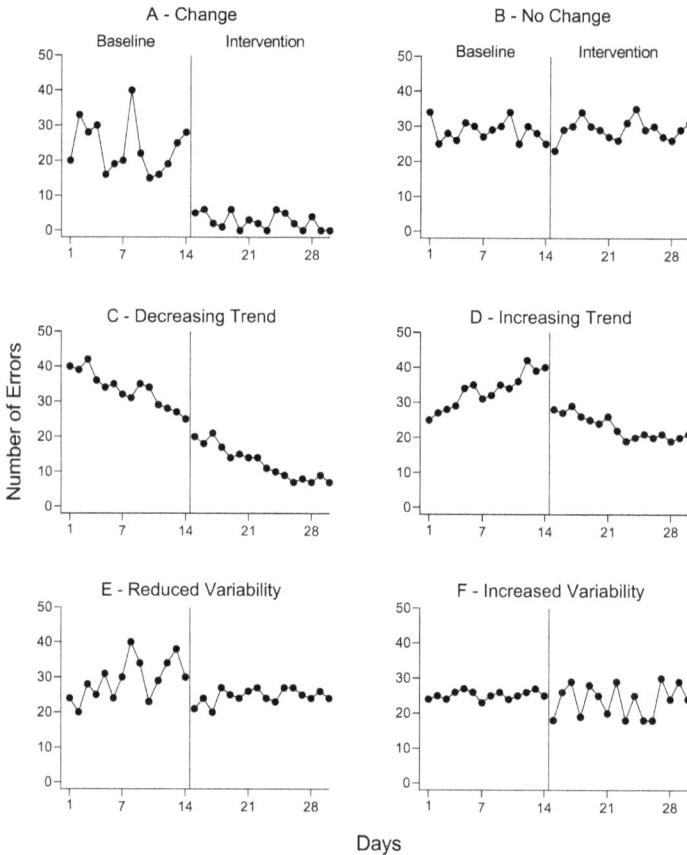

Figure 4.1   Analyzing Change

Number of errors on the y-axis and consecutive days on the x-axis. Data points indicate the number of errors per day. Phase line indicates the introduction of an intervention.

decrease in the number of errors and a reduction in variability. In Graph B, there was no observed level change or change in variability.

In Graph C, change was observed, but might not be due to the intervention because there was already a decreasing trend in the number of errors made during baseline. In Graph D, change could be attributed to the intervention because the trend during baseline was going a different way than after the intervention. In Graph E, the average number of errors didn't change much, but we believe something significant happened nonetheless because of a reduction in variability. In this case, the intervention resulted in more predictable errors and likely influencing variables causing swings in errors. In Graph F, change was observed but might be undesirable. Prior to the intervention, errors were high but not variable. Following the intervention, although there were some days with fewer errors than before, there were also some days with about the same or higher. These increases in variability observed following the intervention are undesirable, suggesting the company no longer has control over what is causing errors.

needed for the remainder of the population who do not respond to group contingencies (Geller et al., 1990). To test this, Ludwig et al. (2010) provided pizza delivery drivers with group feedback. These drivers doubled their turn signal use. When individual feedback was posted along with the group averages, turn signal use nearly doubled again over the previous group feedback intervention. Interestingly, however, drivers who increased their turn signal use in response to the group feedback did not show increases during the individual feedback. Instead, the drivers who increased their turn signal use in response to the individual feedback had not been influenced by the previous group feedback. The lesson for behavioral safety is that while group-level interventions like feedback may be associated with some success, a good portion of the population may still be engaging in the at-risk behaviors. More intensive, individualized interventions will be needed to help them avoid injury.

Indeed, feedback is more effective when provided directly to individuals (Daniels & Bailey, 2014). Further, feedback—like most consequences (Ferster & Skinner, 1957)—tends to be more effective when delivered immediately after behavior rather than delayed. New technologies can deliver immediate feedback to individual workers and can thus have a big impact on behavior. This is clear in studies of warehouse selectors doing cyclical tasks (Berger & Ludwig, 2007; Goomas & Ludwig, 2007; Ludwig & Goomas, 2009) and the seating posture of office employees (Yu et al., 2013). Attempts to apply technologies to deliver immediate safety feedback are just now beginning to be tested. Stay tuned!

In behavioral safety programs that employ social feedback systems, observers discuss the performance directly after the behavior occurs (Step 2;

Figure 1.4 Evergreen Model). This allows for the immediate feedback shown to be effective at both adapting at-risk behaviors and reinforcing safe work behaviors (Agnew & Daniels, 2010; Daniels & Bailey, 2014; Geller, 1996; Ludwig et al., 2010). The feedback can come from multiple sources delivered by fellow employees, managers, and even paid dedicated observers (CCBS, 2017a). Most of the studies (Cooper, 2009; Myers et al., 2010) and books (Agnew and Daniels, 2010; Geller 2005b; McSween, 2003) on behavioral safety call for peer-to-peer observation and feedback where trained employees conduct observations and provide feedback to their peers, who have also been trained in the process to avoid confusion and/or conflict.

An analysis of over 112 million safety observations (Predictive Solutions, 2012) found compelling evidence of the efficacy of employee-based observation and feedback, revealing a clear inverse relationship between observation/feedback rates and injuries. As observations and feedback increase, injuries decrease (a finding replicated across other studies: CCBS, 2022b; Cooper, 2009; Cooper et al., 1994; Fellner & Sulzer-Azaroff, 1984; Hagge et al., 2017; Van Houten et al., 1985). Importantly, the relationship only holds when observation and feedback sessions are conducted by employees instead of safety professionals and trained managers. The analysis also found that the probability of injuries decreases when many observers provide feedback (versus a few dedicated observers). However, there is recent evidence demonstrating that a few dedicated and highly trained observers conducting high rates of observations reduces injuries more than many observers conducting fewer observations (Spigener et al., 2022).

5

REINFORCING
ENGAGEMENT

Once initiated, engagement in process behaviors (e.g., pinpointing, observation and feedback, participation on teams) must be reinforced to shape and maintain the integrity of the program and its effectiveness (Step 3; Figure 1.4 Evergreen Model). The ultimate targets of behavioral safety are the physical task behaviors conducted by workers that protect them in the face of hazards. However, another class of behaviors functionally related to engagement in behavioral safety processes is critical for the direct observation of targeted behaviors, the development of new contingencies, and the introduction of system changes instrumental to the effectiveness of these programs. This class of behaviors includes substantial verbal behaviors on the parts of employees and management as they ask questions for clarification, promote programmatics, analyze data, discuss interventions, and evaluate results. However, the class of behaviors most germane to the success of behavioral safety programs is the engagement of the workforce in the direct observation of peer behavior followed by verbal feedback and discussion.

EMPLOYEE PARTICIPATION BEHAVIORS

This was beyond his comfort zone. Ian had been an upholsterer with his company for nearly 15 years. He enjoyed his work and his colleagues; management a little less. But he gladly took on additional roles by being a trainer and spending a stint working with engineers on a new bedding design. When a new safety committee was formed, he volunteered. He had been on one before, but it didn't do anything meaningful. But this one was different: it was to be run by employees (with management support) and seemed to be focused on what really mattered—the risks employees took at work. He was surprised how much he learned about human psychology in training (he wasn't much of a schoolboy). But then, his peers voted him to be the facilitator of the new team and he realized he was about to begin a task that didn't involve cutting and stitching. He was to be a leader tasked with organizing his team to somehow energize the workforce to participate in a program that, on the surface, seemed a bit duplicitous or two faced. The behavioral safety thing was designed to help workers, but was asking them to put their reputations and jobs on the line by reporting on behaviors that could make them look unprofessional, stupid or reckless. The team had long discussions and concluded that no one would agree to do voluntary observations and most wanted observations to be mandatory (once per month) or incentivized with free lunches and swag. Then management came calling, wanting results right now. What had Ian gotten himself into?

DOI: 10.4324/9781003290711-5

43

Participation counts. Literally! It matters and we should count it. Bottom line: the success of a behavioral safety program depends on participation in the observation process (Geller, 2002c). Participation increases the frequency of individualized peer-to-peer feedback. The more observations, the more conversations. Plus, high numbers of observations provide more data for analysis (Cooper, 2006). Cooper (2009) conducted a meta-analysis on the design factors of behavioral safety systems and found the most behavior change occurs with daily or intermittent (two to three times a week) observations, as opposed to weekly observations. And there is abundant evidence in behavioral safety studies on the relationship between participation in behavioral safety programs and reduction in injuries (CCBS, 2022b; Hagge, et al., 2017; Myers et al., 2010; Predictive Solutions, 2012).

However, many companies report challenges in convincing employees to participate in the observation and feedback process, especially at the beginning (Ludwig et al., 2002). This would suggest that initially, the immediate negative consequences of participation may be stronger than the positive consequences (Agnew & Daniels, 2010; Daniels & Bailey, 2014). While safety observations can lead to fewer incidents, injury reduction takes time to be realized and often cannot be directly experienced by the individual worker when deciding whether to participate (Hyten & Ludwig, 2017). Therefore, injury and other safety outcomes—at least at the beginning of a program—are not really effective at reinforcing participation (Agnew & Daniels, 2010; Daniels & Bailey, 2014).

The fact is that the act of doing an observation and providing feedback causes a response cost[SM-5.1] for the worker (Mayer et al., 2019). The loss of time to engage in other activities can punish participation because it may increase the likelihood of missed deadlines and criticism from the boss. This is especially true in high-volume cyclical work, like in a warehouse. Therefore, management must show support by clearly allowing for the time required to participate in observations.

Science Moment 5.1
Response Cost

A famous behavioral consultant named Tom Gilbert famously said: "Behaviors are costly, accomplishments add value." Behaviors are costly because they take effort—especially in safety, where many safe behaviors take time and effort to do and are often cumbersome and—well, silly looking. Therefore, some behaviors involve a greater response cost than others. High response costs will punish that response and folks will avoid doing the behavior. For example, clipping in your fall protection while climbing a ladder only to have to unclip and clip in higher again and again has a high response cost and therefore probably will be avoided.

We must resist the urge to try to list every behavior related to safety on the observation card, along with adding hazard observations, inspections and permit checks that can take several dozen minutes to conduct. Observation schemes with a large number of behavioral categories increase the amount of time it takes to complete the observation. Geller (1996) argued longer versions may be overwhelming for employees, thus discouraging participation in the process. Indeed, the response cost involved in completing long checklists can easily lead to "pencil whipping" (Ludwig, 2014), where the observer notes what they meant to watch in the first place and then makes up the rest. Reducing the amount of time and effort the observation requires reduces the punishing response cost and potentially leads to an increase in quality participation. McSween (1995, 2003) advised shortening the process by only targeting behaviors that directly correlate with safety outcomes, occur frequently, and are easily observed.

Another potential punisher is that employees may expect to experience awkward verbal interactions during feedback. Talking about safety with a peer is not the norm in an industrial facility. Workers don't want to suggest that their colleagues are somehow less professional than they are; and they expect snide replies along with being the topic of gossip ("Look at Matt being the safety fairy"). Matey et al. (2021) demonstrated that negative responses (e.g., scoffing; eye-rolling; responses like "Seriously, I'm sitting in a chair, I am safe"), and even neutral responses (e.g., an unsmiling straight face; saying "Understood" in a monotone) during feedback can punish participation, leading the worker to cease doing observations in the future.

Therefore, observers should practice basic communication skills and shape tactics for delivering constructive feedback. Observers should also be shaped to focus more on the safe behaviors observed, instead of just pointing out the negative (Geller, 1996, 2005b; Ludwig, 2018; McSween, 1995, 2003). Feedback on safe behaviors reinforces the desired pinpoints exhibited by the worker while making the interaction between peers more pleasant and potentially sufficiently reinforcing to do again. Finally, the workforce should also be trained on how to receive feedback (Ehrlich et al., 2020), so that they accept the coaching as helpful and do not react defensively. And we can't wait until we are in the heat of the moment to practice giving and accepting feedback. Practice must be abundant in training as a safe place to experience the process, emotions, and desirable behaviors (Williams, 2010).

Geller (1996) recommended strengthening participation by publicly posting meaningful results for all to see. The mere posting of observation data shows workers that the behavioral safety process is happening and is being monitored and managed. Most programs post the rate of participation (e.g., number of observations), or the percentage of the workforce submitting observations (also known as a "contact rate").

When behavioral safety programs are first initiated, we can expect low participation rates as workers will be hesitant and in "wait-and-see" mode. This is predictable and okay. When the publicly posted data shows that few workers are engaged, a group of hold-outs (not all of them) are likely to jump in out of fear the program, meant for them, will die. A low percentage rate (e.g., 20 observations) can easily double (e.g., 40 observations). When this is posted, the graphics suggest the program is starting to succeed. Eventually, as participation grows through a larger proportion of the population, graphics continue to display increasing participation rates. This shows how posting observation counts can positively reinforce participation for those engaging in the process. Indeed, Dagen et al. (2009) showed 200% to 400% increases in employee participation due to weekly feedback on observation counts.

It is important to note that among the CCBS accredited programs, the highest participation rate we've seen was 60% of the workforce doing an observation in any given month (FUEL, Marathon Illinois Refining Division; CCBS, 2019). So don't stress trying to get everyone to do observations. Generally, 20% of the population will never do an observation (and will probably be vocal about their negative views—a.k.a "CAVE" people, Ludwig 2018). Then there will be a large sample of the worker population that only do observations rarely. Recent data suggests that a core group of frequent observers is enough to have a significant effect on injuries (Spigener et al., 2022). But we need that core group; and that group must be a representative sample of the workforce (generally 30%).

Once we hit the sweet spot, with a consistent group of observers and an inconsistent group of intermittent observers, continuously posting the observation data will be of no use. You've already reinforced those who already participate, and you won't get more observers. Plus, your plateaued data may make the program look like it is getting stale. We need another number that we can raise!

The quality of observations is the next focus of reinforcement. At the beginning of programs (or in programs that have grown stale), observations tend to report nearly all safe behaviors and don't elaborate on what was seen or talked about in comments. Thus, after participation hits your goal, this is the next target for monitoring, posting, and reinforcement.

High-quality observations are important because they allow for more accurate feedback and more impactful analysis and intervention. Quality observations have been defined as how often the observer recorded at-risk behavior (Laske & Ludwig, 2022b), along with the number of detailed comments written by the observer (Dagen et al., 2009). More frequent and detailed comments allow observers to better elaborate on behaviors they witnessed and describe the environment that may have influenced the behavior

(Kretschmer, 2015). Minimal indications of at-risk behavior (i.e., "all safe" observations) could demonstrate fake reporting by the observer (Cooper, 2006). This has come to be known as "pencil whipping" (Ludwig, 2014) or "tick and flick" in Australia. Without high-quality observations, one-on-one feedback may not address the critical risk being taken. And without comments, future analyses may not reflect accurate conditions and environmental contingencies leading to ineffective interventions (more on that in a later chapter).

Ultimately, group feedback charts display the workforce's performance on behavioral pinpoints (Geller, 1996; McSween, 1995, 2003). This is the whole point! When we have quality observations, we should see some pinpoints with lower safe behavior percentages. This should galvanize the workforce or work crew to improve. As at-risk behaviors are engaged with observation and feedback, percentage-safe should increase over time, eventually meeting a goal. Displays of upward behavioral trends show workers and management that the behavioral program is successfully reducing risk. These upward trends will then reinforce participation in peer observation and feedback, because workers see their engagement in observations and feedback is leading to improvements (Geller, 2002c; Vroom, 1964).

MANAGER BEHAVIORS

Management plays a key role in reinforcing employee participation in the behavioral safety process. Without the involvement of management, behavioral safety initiatives will likely fail (Geller, 1996; McSween, 1995, 2003). This is because of the interlock[SM-5.2] between managers and their workers (Ludwig, 2017b), where managers build contingencies (systems, processes, and rules) which workers behave within. Employees also engage in this interlock through their own behavior, which in turn reinforces or punishes management actions with their performance. Because most organizational systems have intentionally designed this interlock, managers are in a unique position to impact engagement in the behavioral safety process and influence its success in reducing injury.

Employees need guidance and direction at first, or may see observations as "just another unenforced rule" from management and will likely stop participating in the process. Cooper (2006) examined the effect of managerial support on overall safety performance, such as percentage of safe behaviors. He found that managerial support positively correlated with the overall percentage-safe score of the group. But what do we mean by management "support?" We are talking about management behavior that creates the contingencies (management systems and verbalizations) that promote and reinforce employee engagement. Since we are talking behavior here—not some undefined and ambiguous

Science Moment 5.2
Interlocking Behavior Contingencies

A person's behavior does not occur in isolation. It is often "interlocked" with the contingency of others. An "interlocking behavioral contingency" (IBC) occurs when the behavior of one worker interacts with that of another worker and vice versa. These interlocks change the behavioral contingencies for the behaviors of both persons. They can involve multiple people, creating what some call "cultural behaviors." Consider an example from research on nurse absenteeism (Camden & Ludwig, 2013). The behavior of a nurse who calls in "sick" interlocks as an antecedent for another nurse, who now gets called in on her day off. The nurse who is staying home therefore creates a negative consequence for the nurse who has to come in on her day off. This consequence makes the second nurse more likely to call in "sick" herself to skip work in the future, resulting in further interlocks across the rest of the nursing staff, who then start calling in "sick" as well. The cumulative effect of all these interlocks leads to understaffing, additional work, increased patient load, and fewer breaks, further producing adverse consequences for coming to work and thereby negatively reinforcing calling in "sick." One person's behavioral contingency does not occur in isolation and can impact a whole work group. In this example, the influence of the interlock was negative, resulting in lower patient treatment quality, staff fatigue, and higher turnover.

Processes that create safe work environments often involve behaviors interlocked within and between people (and organizational functions). Other people doing other behaviors can greatly impact behaviors proximal to a worker. These other people are working within their own contingencies in other organizational functions distal to workers doing their task (e.g., engineering, planning, procurement, HR, management). For example, a worker who has a broken part on their equipment is now reliant on the maintenance worker who will come to repair the part. If the behavior of the maintenance worker is directed by the supervisor for a different, higher-priority fix, this interlocks with the first worker, in that the equipment is likely to remain broken and the worker will continue to work at-risk.

notion of "support"—we use the same scientific principles we apply to employee behavior. Management behavior needs adequate pinpointing, practice, and feedback to adequately support employee engagement.

Management behaviors are not as algorithmic and physical as those engaged in by workers doing their manual tasks. Instead, management behaviors are typically verbal behaviors[SM-5.3] that can be described as more heuristic. Management utterances, during meetings with themselves or with workers, strategically modify worker contingencies with new or adapted systems, processes, procedures, rules, and other management tactics. More specifically, managers deliver contingencies directly to workers via task

Science Moment 5.3
Verbal Behavior

Talking is a behavior and through our talking, we can influence others' behavior (or our own). Verbal behavior is "behavior reinforced through the mediation of other persons" (Skinner, 1957, p. 2). Skinner further clarified the definition as follows: "Verbal behavior is shaped and sustained by a verbal environment—by people who respond to behavior in certain ways because of the practices of the group of which they are members" (Skinner, 1957, p. 226). In other words, a listener can reinforce or punish the speaker's behavior in an interlock. In fact, verbal behavior is a large part of how behavior in the workplace is influenced, as workers often experience supervision instruction and verbal consequences, as well as peer coaching to shape their behavior. Perhaps the most pertinent for decision making in safety is rule-governed behavior, which is a type of covert verbal behavior (talking to yourself about contingencies—see Science Moment 3.3).

instructions (antecedents) and appraisal feedback (consequences) that are tied indirectly to job outcomes.

Too often, "manager support" is defined merely as getting managers to talk publicly about the program or specific pinpointed behaviors with employees. This tactic relies on exhortations (verbal encouragement), which are antecedents that might direct attention to the safe behaviors but are not sustainable. Otherwise, a common—albeit misinformed—management tactic is to mandate participation by setting quotas and/or offering incentives (e.g., lotteries for prizes, Boyce & Geller, 2001; team prizes for meeting participation goals, Bumstead & Boyce, 2005). Both methods increase participation.

However, employees can meet managerial quota/incentive contingencies simply by turning in observation forms without conducting the observation itself or having the feedback conversation (a.k.a. "pencil whipping"; Ludwig, 2014). In many cases, the observations submitted under managerial contingencies often reflect a bias on the part of the workers, promoting a favorable impression among management (Giacalone & Rosenfeld, 1986) by turning in data showing all their work being done safely.

We recently analyzed 10 years of behavioral data from an established behavioral safety program to assess the impact of a quota system introduced by management in six departments at a chemical manufacturing company (Laske & Ludwig, 2022c). Before the quota, observers had been reporting at-risk behavior in approximately 2.5% (range 0.1–4) of observations a month across all departments. Immediately after the quota was inserted, less than 0.5% of the observations indicated any risk (most observations reported 100% safe). This "near-perfect" rating continued for seven years during the quota scheme and presumably did little to identify and mitigate risk. Similarly, Ludwig (2014) reported on a case where incentives were used to encourage participation where your observation card earned playing cards toward a Texas hold-'em hand which could win a prize. Observations were flat until

Science Moment 5.4
Fixed Interval Reinforcement

"Fixed interval reinforcement" tends to increase behavior only as a deadline approaches. In one experiment by B.F. Skinner, rats pressed a lever to get food, but the food reinforcement only happened after a fixed interval of time. Given that food only became available after a set period of time rather than immediately, the rat's lever pressing would often start slowly and then increase drastically toward the end of the interval. Quota systems around observations counts generate similar "scalloped" patterns of response. Figure 5.1 sets out a cumulative record of observations conducted over a three-month period in a program with a quota system. For every observation conducted, the line increases one point. A steeper slope indicates a greater rate of response, whereas a flat slope indicates a low rate of response. The dashed gray line indicates what stable responding would look like over time. In this example, observations were conducted at a slow rate and were almost stagnant by the middle of the month. However, as the end of the month approached, the rate of observations increased drastically. In fact, 63% of the observations were entered in the last week of the month. Quota systems around observations commonly create these patterns of response.

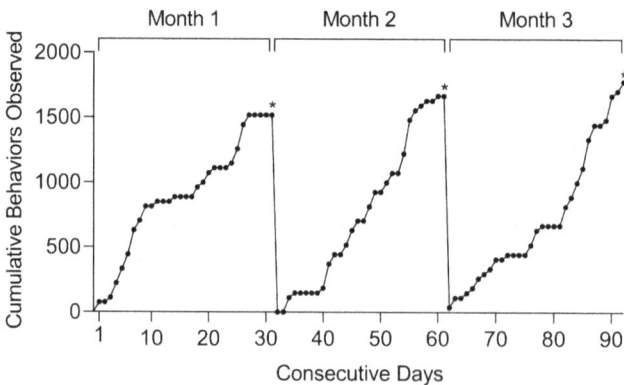

Figure 5.1   Scalloped Response

Number of observation cards turned in on the y-axis and consecutive days on the x-axis. The dark line indicates the number of behaviors observed. The asterisk indicates the quota deadline for each month. The dashed gray line indicates a regression line of 1.0 if behaviors observed were to increase at a constant rate.

the end of the month, when they scalloped upward as the end of the fixed interval[SM-5.4] approached. Each one of the observation cards turned in during the final week was marked "100% safe."

Further, with quotas and incentives, workers are only doing the observations because of the quota and/or incentive. There is a large amount of psychological research on "extrinsic versus intrinsic" motivation (Deci & Ryan, 1985; Ryan & Deci, 2000).[1] That research suggests that if you have a behavior that is done because the person has an interest, engaging in that behavior can be self-reinforcing (i.e., intrinsic motivation). However, once you externally incentivize that behavior (i.e., extrinsic motivation), the motivation for the behavior switches from an internal reason to an external reason. Inevitably, the quota or incentive is the only reason to engage; and when this external reason is taken away, participation drops. The intrinsic motivation, as psychology researchers describe it, has been "undermined." This means voluntary participation will be mighty hard to get back.

You can see this over a nine-year period in an examination of observation data from St. Paul Park Refinery's "All Work At-Risk Eliminated" (AWARE) accredited program (Figure 5.2) before it became successful. This figure shows

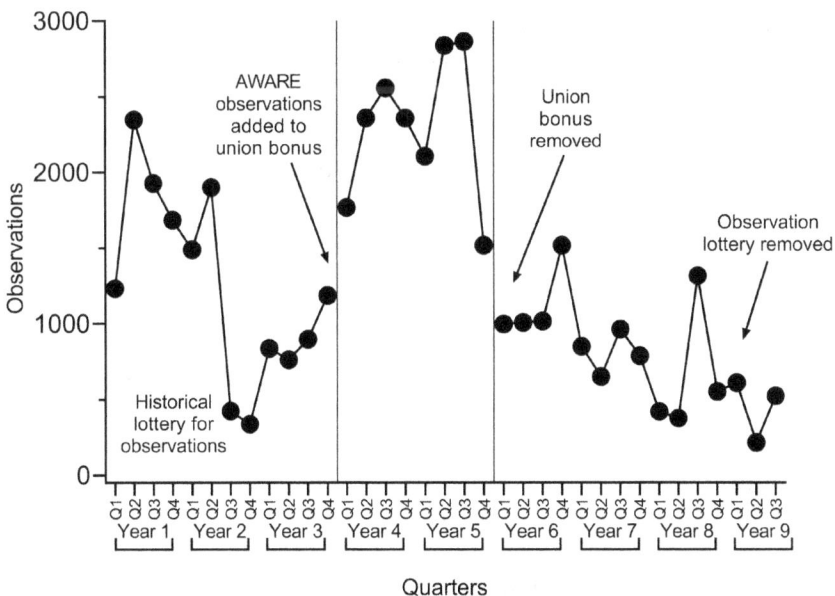

Figure 5.2   CCBS Accredited Program AWARE: ABA Design.

Number of observations on the y-axis and consecutive quarters on the x-axis. Data points indicate the number of observations per quarter. Data were adapted with permission from the CCBS. From "Behavioral Safety: An Efficacious Application of Applied Behavior Analysis to Reduce Human Suffering," by T.D. Ludwig and M.M. Laske (2022) *Journal of Organizational Behavior Management* (Taylor & Francis, 2022).

the impact of making union bonuses partially contingent on submitting observations over a three-year period (CCBS, 2020b).

A previously high voluntary participation rate dropped off in Year 2—probably because of some event that upset the program (e.g., new leadership, a company merger, a production boom, or an emergency). Management instituted a lottery incentive (which can be used effectively; Boyce & Geller, 2001; Geller, 1984; Geller et al., 1989; Ludwig & Geller, 1991; Nimmer & Geller, 1988; Rudd & Geller, 1985). This lottery was effective in gradually returning participation to its previous level over the next year. Unfortunately, however, participation in observations was then incentivized and linked to the union bonus. While this resulted in the predicted high levels of participation, it also produced low-quality observations which did not mitigate injuries. When taken away after two years, participation fell. Ultimately, the lottery was dropped and the participation rate after nine years of tinkering with incentives ended up substantially lower than when the program began with voluntary participation.

Instead of mandates and incentives, we should instead pinpoint the manager behaviors that reinforce the behavioral safety program by adapting management systems (Cooper, 2006) within their ongoing operations. This may involve changing workflows (e.g., standard operating procedures), administrative processes (e.g., scheduling, requisitions), policies, reporting structures, functional groupings of effort (e.g., maintenance, operations, engineering), external support (e.g., consultants), and company infrastructure, technologies, measurements, and communications to make participation in behavioral safety reinforcing instead of punishing. The primary role of management in behavioral safety is to learn from behavioral observation trends to build and maintain contingencies that support behavioral safety programming and execution.

Here are some examples. A primary management role is to create and train an employee team to manage the behavioral safety program. Managers secure the labor time for the participants, coordinate scheduling, and assure adequate training and facilitation. Manager behaviors should involve monitoring and reacting to key process measures that evaluate the maturity and health of the behavioral safety program (Cooper, 2006; Zohar & Luria, 2003). Manager behavior pinpoints are important when we are analyzing behavioral data coordinating different functions (e.g., engineering, maintenance, procurement) to adapt systems, equipment, and workflows to mitigate risk found during observations. Finally, managers can adapt their communications processes to bring behavioral data into decision making in ongoing production meetings.

Managers are human and their behaviors are selected by consequences, the same as other employees participating in behavioral safety. We need to pinpoint the behaviors managers need to engage in. We need to monitor these behaviors and reinforce them with feedback. Sound familiar? Specific pinpoints will be idiosyncratic to the industry and company; but pinpointing and subsequent monitoring and feedback on manager behaviors are essential and ongoing activities in any behavioral safety program.

Note

1   Much of the work in psychology around intrinsic versus extrinsic motivation has been disputed in behavior analysis. A review of both perspectives is beyond the scope of this book. We direct you to Dickinson (1989) for an exceptional review of the empirical evidence and the behavior analytic interpretation.

6

TRENDING **AND** FUNCTIONAL **ANALYSIS**

There had been a run of leg injuries at the warehouse in recent years. One primary cause was selectors stepping off their moving hand truck and getting hit or pinned. Because of this, the behavioral safety team pinpointed "Stopping hand truck fully before stepping off" as one of their targets. They knew this was an obvious area of risk and baseline observations verified that selectors were only stopping fully 50% of the time (and this was probably high). Over the next couple of months, peer and group feedback on complete stopping raised this to about 63%; with further emphasis, this topped out at around 80% for a couple of months before a subsequent downward trend. This was a "stubborn" behavior which trends suggested could not be fully improved through feedback and shaping. A more sustainable solution was needed. The team dedicated a meeting to conducting an Antecedent-Behavior-Consequences (ABC) analysis on this pinpoint and invited leaders and engineers to help. One conclusion concerned the configuration of braking on the hand trucks. After slowing, the hand truck could be put into a gear that would maintain some momentum before stopping. Selectors liked this because they could step off the moving vehicle, pick their case, and when they turned around, the pallet at the back of the hand truck would be right in front of them: no walking needed to set the case. Thus, stepping off the moving hand truck was reinforced. The only real sustainable solution was either to remove this gear or replace the hand trucks. Everyone knew this was a problem, but the cost of fixing all the hand trucks was an issue—as was the perceived productivity loss of fully stopping. In the end, the behavioral data helped convince leaders that fixing the hand truck problem was worth the investment because they could see how the configuration was leading to risk and injuries.

Observations also produce valuable data! Certainly, information gained during observations leads to effective peer-to-peer interactions on critical behaviors which have a big impact. However, consider the value of the data collected across the plant, and how it gives insight into the real risks that may be happening—risks whose magnitude managers may be unaware of (Ludwig, 2018). When large amounts of data are collected, trends in at-risk behavior emerge. These trends showing risk can then help us target the at-risk behavior for further analysis with the goal of designing interventions that create sustainable behavior change. Trending and functional analyses are conducted (Step 4; Figure 1.4 Evergreen Model) when pinpointed behavior does not respond to direct feedback.

DOI: 10.4324/9781003290711-6

It is hard to trend injuries with any real success because, thankfully, they don't occur very often. Greater attention has recently been paid to serious injuries and fatalities (SIFs) (Chi et al., 2014; Chiang et al., 2018; McSween & Moran, 2017), as well as process safety incidents (PSIs) (Bogard et al., 2015; Gravina et al., 2017; Hyten & Ludwig, 2017; Lebbon & Sigurdsson, 2017; Ludwig, 2017b, 2017a), which—very thankfully—have a very low base rate. A company's injury rate is defined by regulatory reporting standards (BLS, 2019a), which don't account for minor injuries and first aid; these don't need to be reported, because they don't require medical attention, prescriptions, or days off work. Companies can—and should—gauge their safety performance on these outcome measures. However, using recordable injuries as the only basis for trending and analysis relies on reactive action to a "lagging indicator." The reasons for the incident have already happened and someone got hurt. While fixing these are important, they do not help predict where the next injury will occur.

In response, the safety profession adopted "leading indicators" as a source of more abundant data that can be used to proactively analyze and intervene to prevent injuries before they happen (OSHA, 2019). Common leading indicators include reports of minor injuries/first aid and close calls where energy is released but doesn't impact a body. Leading indicators can also include operational measures such as preventive maintenance, equipment damage, inspections and audits, and hazard identification. All these measures offer a higher base rate but present something of a contraindication for management. Managers may see high levels in these leading indicators as a bad thing, suggesting a higher probability of injury. The paradox is that greater reporting of leading indicators is preferred because higher rates of these reports provide clearer insight into the reality of hazards and risks in the workplace (Geller, 1996; McSween & Moran, 2017).

While leading indicators provide better data that helps identify safety issues to mitigate, they still leave a gap. An event must happen, and a worker must feel comfortable enough to report it (Ludwig, 2018). Behavioral data, collected through direct observation, can fill this gap because it reveals risks currently engaged in by the workforce. After all, most lagging and leading indicators are directly related to behavior. Injuries, close calls, equipment damage, and the like are the product of behaviors interacting with their environments. Behavioral data comes from direct observation, while the behavior is happening, in the context of the environment. This is a much richer palette of information upon which to analyze and make decisions to target future behavior.

TRENDING BEHAVIORAL DATA

Behaviors either happen or they don't. We don't pinpoint behaviors so that we have to use subjective judgments on a scale of one to five. The behavior is recorded as having happened or not, giving us response options of "safe" or "at-risk." To trend the observational data, we then aggregate the binary data into a percentage of observations of safe behavior (i.e., percentage safe) per pinpoint category for a period of time.

Here is an example of some averages based on hundreds of worker observations over a month from the Advantage Logistics Southeast Regional Facility's Critical Activities Management (CAM) program (Figure 6.1; CCBS, 2009). These averages direct attention to at-risk behavior trends around using forklifts and low case hooks—a safe behavior when pulling cases from the back of large shelving. It looks like wearing proper personal protective equipment was going pretty well this month.

As we've discussed, the goal of any new pinpoint is to identify risk when added to the observation protocol. A new pinpoint showing nearly 100% safe behaviors, while good news, may be a product of pencil whipping (requiring a return to Step 3; Figure 1.4 Evergreen Model). Otherwise, if the safe behavior is indeed near 100%, we may be spending resources on observing behaviors that the workforce has already mastered. If this is the case, we need

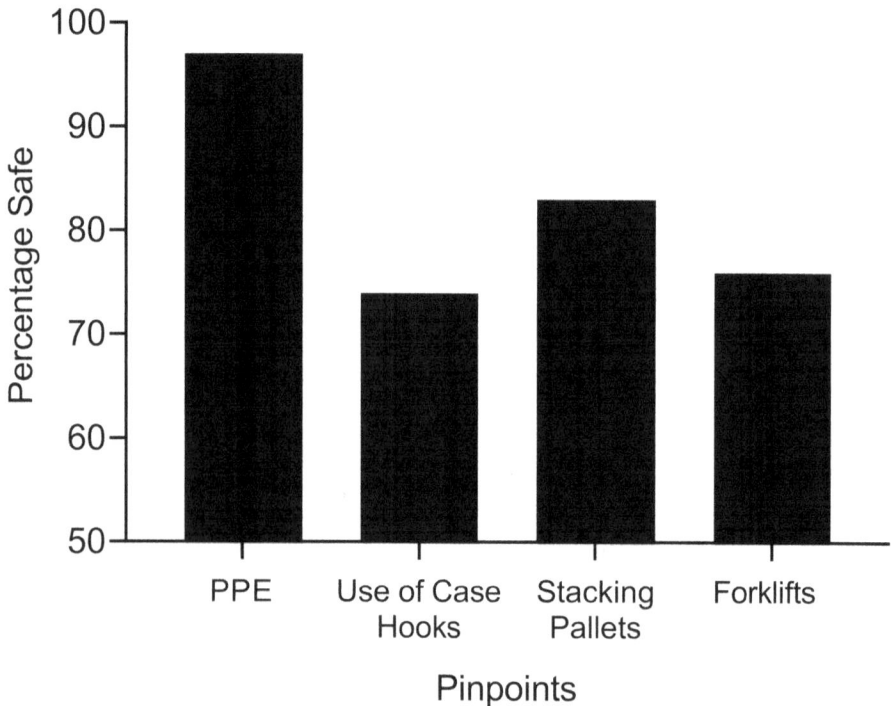

Figure 6.1   CAM Program Percentage of Safe Behaviors by Pinpoint Target

Percentage safe behaviors are on the y-axis and pinpoint targets in each bar on the x-axis. Data is adapted with permission from the CCBS.

to return to Step 1 (Figure 1.4 Evergreen Model) and adopt a new pinpoint to find a trend of at-risk behavior that represents the true state of affairs in the workplace.

The objective is not to find instances of at-risk behavior and then immediately make changes; or worse, hunt down the "perpetrator" to try to fix this person via training or discipline. Behaviors vary from environment to environment and over time. When single observation finds risk, it does not mean this risk is prolific among the workforce or even for that single worker. The role of trending is to verify that the behavioral risk is well sampled and not just a product of random variation. When we find a trend of at-risk behavior, we know that the behavior is a product of the workers' environment instead of something to do with an individual worker (otherwise, we wouldn't see it across the board).

Trending is the first step in a process aimed at determining the causes of at-risk behavior, which may be obvious or complex. For example, some job tasks are cyclical and under time pressure. Trends showing only 50% safe behaviors in these tasks suggest significant behavioral risk related to the design of the work or management systems governing the work (e.g., work standards or quota systems).

An example of trending is shown in Figure 6.2 from a behavioral observation program at a grocery distribution warehouse. Presented in the figure are six months of data trends on four new pinpoints. Each of the pinpoints, when observed, was successful in finding behavioral risks, as each started with baseline observations well below 80% safe. This was a victory, as these trends empirically demonstrated the risks present in the workplace and confirmed these were areas where workers needed help.

When the employee team adopted the safe pinpoint "Stop equipment before stepping off," they had done their homework by looking at injury data showing strikes to feet and legs from still-moving equipment they had just stepped off. Their baseline observations confirmed this to be a problem, showing just a tad above 50% complete stops. Over the next six months of observations and peer feedback, together with an awareness campaign in June and July, they were unable to successfully increase this behavior up to their goal levels of 90%. This trend suggests that workers "trying harder" and getting feedback was not sufficient to change the behavior. Instead, there was something in the workers' environment maintaining the at-risk behavior. Indeed, as the team engaged in ABC analysis (discussed in the next section), they noted that the hand truck equipment was designed to allow the worker to coast to a stop instead of brake to a full stop. This was advantageous for production, in that the worker could pick their cases off the stacks and, when they turned around, the back of the hand truck would have coasted right to where the pallet was at their feet, ready for the worker to put down the case. The behavioral safety team was able to demonstrate the

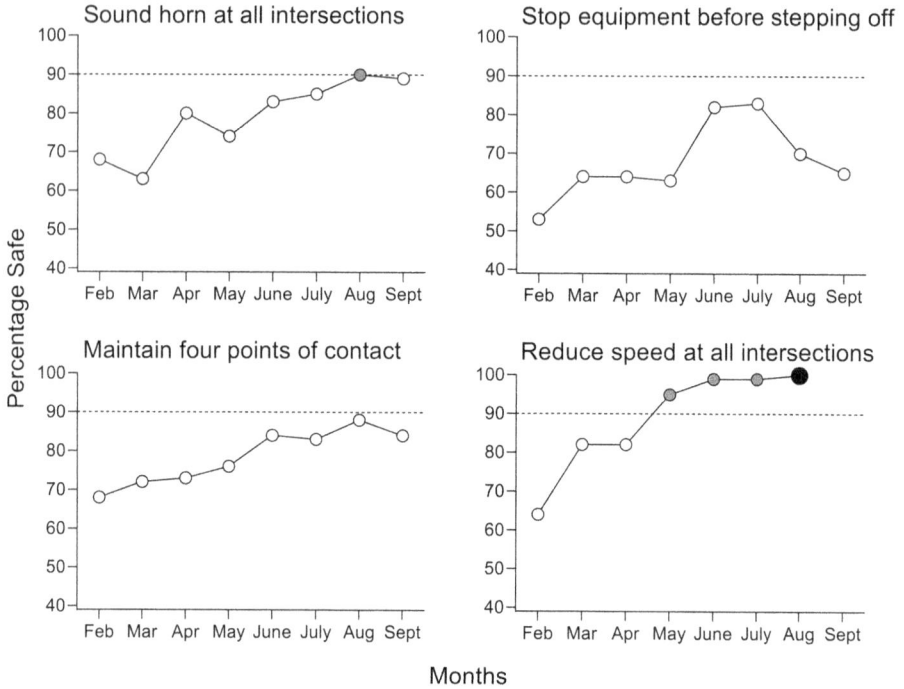

Figure 6.2   Grocery Distribution Pinpoints

Percentage safe behavior on the y-axis and consecutive months on the x-axis. Gray circles indicate performance above the goal. The black circle represents when the checklist item met the three consecutive month criterion and was retired from the checklist.

magnitude of the risks being taken related to recent injuries, which allowed them to enter into data-based discussions with their operations leaders to make changes to the equipment.

You can get really fancy and correlate monthly behavior trends with production numbers or quality indicators. Correlations related to demographics such as product division or shift, environmental conditions or other pinpoints may lead to interesting results and target the best areas for improvement. Our research lab has been involved in analytics recently in a bid to empirically show some of these relationships. Laske et al. (2022) found relationships between shift changes (night to day) and the likelihood of near-misses and injury. Hinson and Laske (2020) found a higher risk of injury on windy days. Ludwig & Geller (1991, 1997) found that turn signals tended not to be used when the driver did not come to a complete stop.

Of course, most behavioral safety teams and safety professionals are more likely to use "low-tech" options for data trending. All that is really needed is a well-designed MS Excel file or integration into company IT that contains the same spreadsheet capability where you can do some simple math and track data over time. Certainly, you can spend more money on some fancy data capture technology; but our experience is that you have limited ability to really play with the data (unless you pay for upgrades).

Once you have your trends, they should be reviewed and analyzed within a group process. A group comprised of people familiar with the job and the different systems affecting worker behavior should be called upon in the review to provide their experience and expertise to suggest potential causes of the at-risk behavioral trend. These include workers who do the tasks and are familiar with the environmental conditions in which they take place. Others who know the management systems involved in the task (e.g., training, scheduling, permitting), the state of the equipment (e.g., maintenance), engineering specifics, tool procurement, human resources, supervision, and even financial experts should be invited to take part in analyzing these trends, to help identify system variables in need of intervention.

FUNCTIONAL ANALYSIS

If we can effectively and accurately analyze at-risk behavior trends, we can sustainably reduce injuries by directing interventions toward factors that put workers in a position to take risks. At-risk behaviors are preceded by a plethora of antecedents in the workers' environment (e.g., hazards, tools and equipment, facilities, supervision, peers); system factors (e.g., onboarding and training, process design, incident reporting, policies and procedures); behavioral interlocks related to organizational functions (e.g., procurement, human resources, operations/production, engineering); and market realities known as metacontingencies[SM-6.1] (Glenn, 1988; Glenn & Malott, 2004) that impact decision making (Ludwig, 2017b).

While this complex web of interacting causes may be difficult to unravel, all impact the behavior of frontline workers, whose actions in the face of hazards put them at risk of injury. For example, a worker may have a production schedule to meet, but their equipment may be malfunctioning. The worker may be forced to engage in short-term fixes and workarounds, and open themselves up to injury. Further, the worker's behavior can be affected by organizational-level metacontingencies. This complex interplay of interlocking contingencies forces us to consider the multiplicity of functions that make the behavioral systems functional in creating the environmental contingencies that impact frontline workers (Agnew & Uhl, 2021; Malott, 2003).

Functional analyses in behavioral safety analyze pinpointed behaviors for proximal contingencies in the immediate work environment, as well as the more distal "systemic" factors that impact the work environment (Gravina, Nastasi, & Austin, 2021). The proximal and distal environments work together. Proximal solutions—such as ensuring workers have proper tools at hand for the task—depend on distal systems, such as a warehouse that keeps the

Science Moment 6.1
Metacontingencies

A "metacontingency" considers all the interlocking behavioral contingencies (IBCs) and how they work together to achieve something (good or bad). Consider that accomplishment—which is the product of a whole bunch of different people's IBCs—as the metacontingency's version of behavior. Companies have metacontingencies; as do their different functions, workgroups and crews. The metacontingency has its own antecedents, but these typically are the things in the business environment that direct the behavior of everyone (e.g., a competitor's new product or a new government regulation). Metacontingencies also have their version of consequences. The accomplished thing (e.g., product, resource, information) is then often used by a customer (internal or external to the organization) to see if it helps them and adds value to their world. The customer provides something back in return. In a business environment, this may be revenue. Internal customers may provide feedback or their success may impact bonuses. Therefore, internal and external customer demand is likely to shape the interlocking behaviors within the organization.

proper tools in stock, to be sustained. Proximal contingencies are identified using the ABC analysis (Daniels & Bailey, 2014; Geller, 2005a; Gravina et al., 2014; McSween, 2003; Petrock, 1978; Sulzer-Azaroff, 1987); while more distal interlocking contingencies are unraveled using principles from behavioral systems analysis (Diener et al., 2009; Ludwig & Houmanfar, 2010; McGee & Crowley-Koch, 2021).

The functional analysis of safety behaviors takes place over four steps:

- direct observation of behavior and the environment;
- pinpointing and ABC analysis;
- process analysis; and
- behavioral systems analysis.

These steps may be iterative, in that discoveries in one step may take the analysis back to previous steps as new pinpoints and contingencies are discovered.

Direct Observation

When discovering an at-risk behavior trend, the first step is to assemble a group to conduct direct observation of the tasks in which the behaviors occur. This group should be representative of employees who do tasks associated with the behaviors and of direct supervision and management, along with specialists in, for example, engineering (to consider design and materials issues); maintenance (to consider hazard mitigation); human resources (to consider scheduling and shift work); and quality and LEAN (to consider workflow)—as well as the obvious presence of safety professionals. It is imperative that these individuals travel to the workplace and directly observe tasks that involve the behavior.

As they witness workers engaging in the task, analysts take note of:

- the topography of behavior (i.e., what the behavior looks like);
- competing behaviors that are done instead of the safe behavior;
- the association of the behavior with hazards in the environment;
- the workflows of behaviors preceding and following the pinpoint;
- direct interlocks with fellow workers as they pass off materials, information and equipment from one task to another;
- direct antecedents (e.g., permits, signage, supervisory instructions, prompts from fellow workers);
- direct consequences (e.g., time and effort to complete a task, accessibility of next task); and
- other observable factors that potentially relate to the at-risk behavior.

Pinpointing and ABC Analysis

In a subsequent meeting session, the analyst group clarifies the behaviors related to the safe performance of the task(s) along with the competing behaviors related to risks. Note that the behaviors ultimately analyzed may be different from the pinpoints on the behavior observation cards. Often, direct observation of the environment can lead to more detailed and/or appropriate pinpoints for the situation.

The group then performs an ABC analysis (Geller, 2005a; Gravina et al., 2014; McSween, 2003; Sulzer-Azaroff, 1987) of direct-acting contingencies setting the occasion for the behavior. (Full descriptions of ABC analysis in behavioral safety can be found in Daniels, 1989; Daniels & Bailey, 2014; McSween, 2003; Petrock, 1978.)

In an ABC analysis, the analysis group first considers the consequences of the safe behavior. Common consequences tend to be "avoid injury and/or discipline" as negative reinforcers and "extra effort, discomfort, cumbersome" as natural punishers of safe behavior. Consequences are then assessed based on their probability, how prompt it happens, and if the consequence is personally desired or undesired for the worker. This determines the strength of the consequence.

Most likely, negative reinforcers of safe behavior are not very powerful because injury is rarely the result of engaging in the variant at-risk behavior (i.e., it is not "probable"). Often, the consequence of the safe behavior—"avoiding injury"—can never be realized. How does one experience something that hasn't happened? In contrast, punishers of safe behaviors are more probable, immediate, and desirable; this often explains the omission of safe behaviors and the presence of at-risk behaviors.

Antecedents are listed next. You'll find that these are substantially more abundant. Many antecedents are not linked to consequences (e.g., signs, training), but can prompt behavior if done well. But typically, prompts only impact behavior for a short time until behaviors drift back to the environmental contingencies present. The more powerful antecedents are those that are in the context of the work and have a clear connection to the consequences—especially the consequences that ultimately punish the safe behavior. For example, an antecedent analysis may suggest that workers often find themselves without the right tool for the job. This is a powerful antecedent because it relates to consequences such as delayed production because the worker must search to find the right tool. We can conclude that the at-risk behavior is more likely to happen when the correct tool is not immediately available.

Table 6.1 depicts a completed ABC analysis for a grocery distribution warehouse. You can see how the at-risk behavior—stepping in front of a coasting pallet jack (equipment driven by selectors)—is reinforced by the consequences, so selectors are motivated to take the risk. A look at the antecedents shows how the equipment, processes, production pressure, culture, and even the warehouse's safety management systems put the worker in a position to take the risk.

When a good ABC analysis is conducted, we often come up with fairly straightforward solutions. For example, a worker is observed using the wrong-sized clamp when installing a coupling on an oilfield hose. The risk is that the coupling may fail and explode when a future employee is nearby. Analysis of this consequence determines that the avoidance of future disaster is not a probable, prompt, or personal negative reinforcer. Instead, our ABC analysis describes immediate antecedents around the worker's clamp choice. The correct-sized clamp is back at the warehouse (antecedent), a 20-minute walk away (punishing consequence). Use of the available clamp is negatively reinforced by avoiding the walk and avoiding a supervisor questioning the worker's absence and lack of timeliness on task completion. This analysis then leads us to build a process through which tool cribs have the right tools for the day's tasks. The next step in the functional analysis in behavioral safety addresses the broader processes that created this environment that caused the risk (i.e., an inadequate tool crib).

Table 6.2 shows a solution analysis using the ABC format. These solutions relate to the issues discovered in the ABC analysis shown in Table 6.1. The team identifies the safe behavior as throttling down or stopping the pallet jack before crossing. Note how the team suggests further analyses of consequences to determine the real time saved versus the probability of a serious incident occurring. Then, the antecedent analysis informs interventions in each of the areas to direct selectors toward the safe behavior.

Table 6.1 ABC Analysis for Grocery Distribution Warehouse

ANTECEDENTS

EQUIPMENT
Coast pro slows jack down below production speed.
Jack designed so selector can step off toward the front.

PROCESS
"Z-picking" requires crossing in front for next pick.
When on direct pick, selectors don't know where the next slot is, so they walk the jack.

PRODUCTION PRESSURE
Production contributes toward bonus.
Leaders and supervisors encourage production.
Prior tickets, congestion, and other events slow production and time must be made up.
No overtime, so people hurry to complete work to go home.

CULTURE
Senior selectors model this risk.

SAFETY MANAGEMENT SYSTEMS
No rules on walking in front of a moving jack.
Training teaches walking in front at "arm's length," but teaches nothing about stopping beforehand.

AT-RISK BEHAVIOR

Coasting the pallet jack while stepping off and stepping in front while jack is still moving.

CONSEQUENCES   *This will happen to me*

Save valuable seconds which add up to earn unlimited bonus and/or praise/gift cards.
At the end of a shift, the volume may require the shift to stay over until complete. Saving time on each pick reduces this extra time.
The jack actually hitting the leg and causing injury is unlikely (or at least perceived as unlikely).
If the jack is still moving, it will pass you and the cargo pallet will be even with you as you pick the case. This way, you don't have to walk any further to place the case on the pallet.

Process Analysis

The ABC analysis looked at the immediate, proximal environment influencing the at-risk behavior. As a first step to understand the events leading up to the at-risk behavior, we need to analyze important events (or non-events, most

Table 6.2 Solution ABC Analysis for Grocery Distribution Warehouse

ANTECEDENTS

EQUIPMENT
Research equipment with center consoles that keep selectors from stepping in front. Pilot and collect data on center console impact on safety and production.

PROCESS
Order picking technology to provide information on the slots ahead so the selector can plan how to route themselves.
Move toward U-pick.

PRODUCTION PRESSURE
Analyze shift data on front walking to determine how much time it takes.

CULTURE
More behavioral observations on this pinpoint to engage more people with feedback.

SAFETY MANAGEMENT SYSTEMS
Incorporate alternatives to walking in front in onboarding jack training.
Create video demonstrating alternatives to walking in front while in motion but still maintaining production.

SAFE BEHAVIOR

Throttle down jack or let jack stop completely before crossing in front at arm's length so it doesn't hit you.

CONSEQUENCES

Analyze shift data on front walking to determine how much time it takes versus the safe target.
Increase near-miss reports. Collect and analyze near-misses and minor injuries and injuries due to walking in front (publicize and discuss during onboarding).

likely) in the worker's day. Here we will often find where our safety management systems have let them down, leading to risk.

Starting at the moment of the at-risk behavior, create a process map of the steps the worker was doing to complete the task. Then build that process map backward by outlining the entire day of the worker—a path that usually includes:

- a pre-shift meeting at which work instructions and risk assessments should be discussed;
- gathering the proper tools and equipment;
- inspecting the job site;

- preparing work (e.g., getting work permits, lockout energy, and tag equipment);
- communicating with others before and during tasks;
- undertaking the task itself (steps already mapped);
- delivering the product of that task for the next person to work with; and
- post-task reporting of deviations (e.g., near misses, damaged tools/ equipment, miscommunications, unexpected events).

We will show an example of a process map[SM-6.2] in Figures 6.4 and 6.5, in the next section on behavioral systems.

For each process step, consider other people and functions involved. Add their processes as additional "swim lanes" (Brache & Rummler, 1997; M. E. Malott, 2003). For example, the pre-shift meeting is typically convened and facilitated by a supervisor who has their own process to determine the day's work, assign tasks and schedules, and prepare for risk analysis. When considering the tools and equipment that workers retrieve, one could consider the procurement department's process of making the right tools and equipment available for the job. The worker's job site inspection will reveal conditions left from the process by the previous shift or by an external contractor who may be working nearby. Each of these other people engages in a process whose quality and outcome impact the workers' environment that provides the contingencies for safe or at-risk behavior. In many cases, we find additional pinpoints (of other people) to target with observation/feedback and analysis. As this analysis continues to reach back into behaviors performed by others in the operations system more distal to the frontline worker, we begin to engage in behavioral systems analysis.

Science Moment 6.2
Process Mapping

"Process mapping" is a method to describe and document the behaviors, steps, and decisions that are required in critical processes. When creating a process map, horizontal "swim lanes" are used to indicate the different people, functions, equipment, or technology involved in the process. Figure 6.3 sets out an example process map for an assembly line. Within the different swim lanes are each assembly worker, the parts inspector, and the forklift operator. Process mapping is a useful tool to identify redundant activities, bottlenecks, excessive reviews, unnecessary steps, and variability. A poorly designed process will often put employees in a position where at-risk behavior is encouraged, and safe behavior punished. Rummler and Brache (2013) summarized the importance of understanding processes as follows: "If you pit a good performer against a bad system, the system will win almost every time" (p. 11).

Assembly Line Process Map

Assembly Associate A	Grab part → Complete step one assembly → Place part on assembly line		
Assembly Associate B		Grab part → Complete step two assembly → Place part on assembly line	
Parts Inspector			Grab part → Inspect part for error → Error in assembly? (Yes / No) → Place part on assembly line
Fork Lift Operator			Grab part → Place part on forklift pallet → Take pallet to warehouse

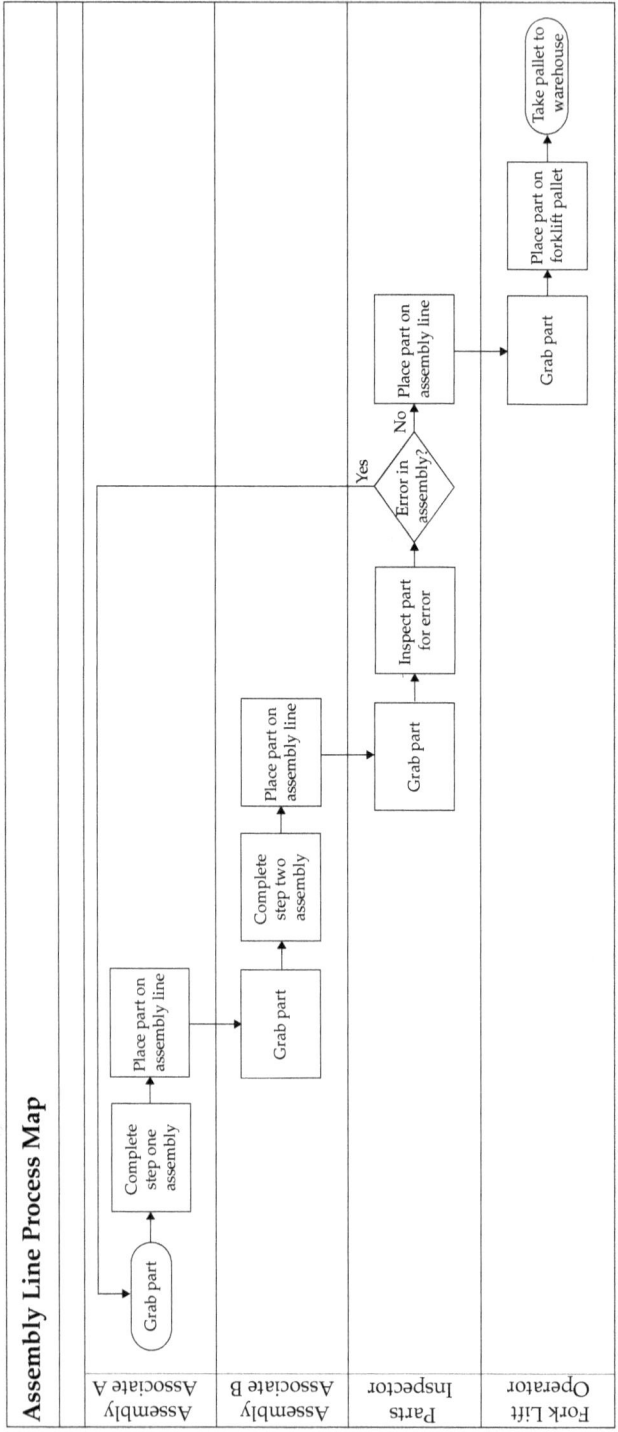

Figure 6.3 Assembly Line Process Map Example

Behavioral Systems Analysis

After completing the process analysis, you will undoubtedly see that certain safety management systems (e.g., pre-shift meetings, permitting, inspection of tools) may have been omitted or pencil whipped. A fuller process analysis with additional "swim lanes" showing what other people were supposed to do in their own processes may reveal why the worker didn't have the proper tools, equipment, instructions, setup, or other critical antecedents to direct their safe behavior. Now you see you have a bigger problem to solve—and you have barely scratched the surface!

The point is that your work processes have been built logically to deliver an environment that promotes safe behaviors. Yes, you've got the pre-shift instructions, hazard observations, permits, inspections, lockout-tagouts, standard operating procedures, and a myriad of safety management systems—often with fancy acronyms—designed to deliver safe performance. However, even logical processes can be disrupted. After all, processes are just a sequential list of behaviors interlocked within and between people (and functions).

Other people doing other behaviors can greatly impact the delivery of safety and other important processes to workers, putting us back where we started. Often these other people doing other behaviors are working under their own contingencies in other organizational functions (e.g., engineering, planning, procurement, HR, management), far away in time and space from the worker. This complex interplay of interlocking contingencies forces us to consider the multiplicity of functions that make up the behavioral systems creating the environmental contingencies that impact frontline workers (Diener et al., 2009; M.E. Malott, 2003).

Let's consider a high-potential close call that was observed by a behavioral safety observer at a refinery (true story). One morning, a forklift driver was delivering a pallet of tools from the warehouse for maintenance workers doing an afternoon repair on a unit in the refinery. Another employee doing a behavioral safety observation saw that the mast of the forklift was about to hit an overhead chemical pipe. The observer intervened by stopping work and a disaster in the form of an explosion, a chemical release and severe injury was averted. Because this was such a serious potential event, the behavioral safety team decided to conduct a functional analysis to determine the causes and plan interventions to prevent similar incidents in future.

They began by clarifying the at-risk behavior (Geller, 2005b). Initial definitions of "Did not look up" or "Did not follow rules" were discarded because these failed to pass the dead person test (Lindsley, 1991) and would have ended up in exhortations (e.g., training sessions teaching people to look up and follow rules). Instead, the active at-risk behavior was defined as "Driving forklift under low piping in restricted area." The next step was to define the safe alternative: "Drive route around outside of equipment."

The ABC analysis described the consequences that explained why this safe behavior didn't happen. The answer was obvious: the restricted pathway within the unit was the fastest route to the equipment drop-off point. The time saved was a direct-acting positive reinforcer.

But what about the antecedents that would have discriminated the safe route? One would think that a driver who knew of the major overhead hazard would not have taken that route. To better understand this particular behavioral environment, the team started mapping out the process the driver followed. Figure 6.4 shows the process map for the delivery of construction tools within a refinery.

Circle A in Figure 6.4 shows the point at which the risk was identified by the observer and work stopped. Going backward from here, we see the driver should have engaged in the refinery's permitting process. A permit is an authorization from the local operators of the equipment that allows others to enter their space safely. The permitting process requires a number of checks to verify the right thing is being delivered, the equipment can operate safely, worker certifications have been obtained, and so on. One part of the permitting process is for the operator to walk the safe route to the delivery point with the driver, pointing out specific hazards and restricted areas. If this had happened, the driver would have experienced the antecedent (i.e., the overhead hazard) that would have discriminated against taking that route. It was evident, therefore, that the operator had not conducted the walkabout with the driver (Circle B in Figure 6.4). While the ABC and process analysis clearly identified why the delivery driver was in a position to take the risk, a new, more impactful at-risk behavior was found.

The analysis continued, but now based on the interlock that caused the original close call. We were no longer looking at the at-risk behavior performed by the driver. Instead, we had a new at-risk behavior to investigate: "Operator conducted a 'couch permit' and signed permit without walking the route." A simple ABC analysis revealed that walking the route punished the operators doing the permit process, and that "couch permit" behaviors were shaped (see Circle B in Figure 6.5). More specifically, the act of walking the route with the driver made the permitting process last an average of 20 minutes. The response effort was potentially punishing in that this walkabout took them away from other priority work or the weather could be bad.

The question then became: "What antecedents were in the operator's world that told them not to do the walkabout with the driver?" In scientific terms: "What discriminated this negative punisher?" The operators participating in the analysis were quick to point out (and had data to prove) that the permit system was out of control (Deming, 1986). They were getting an average of 15 permits daily, varying widely from day to day (see Circle C in Figure 6.5). The high number of permits cumulatively increased the

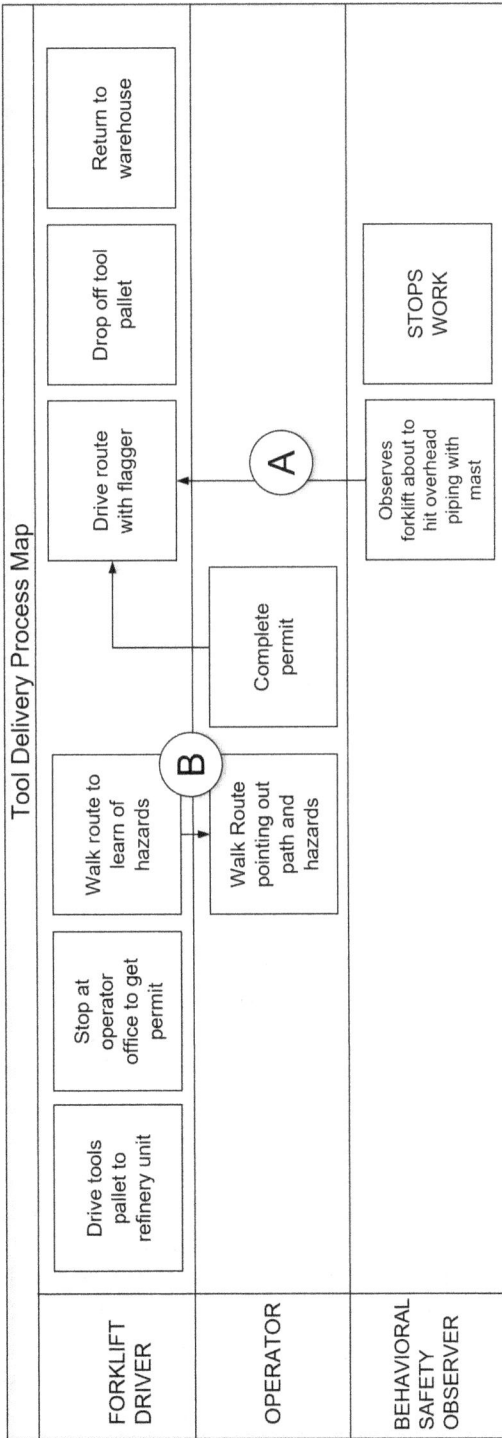

Figure 6.4   Map of Tool Delivery in a Refinery Process

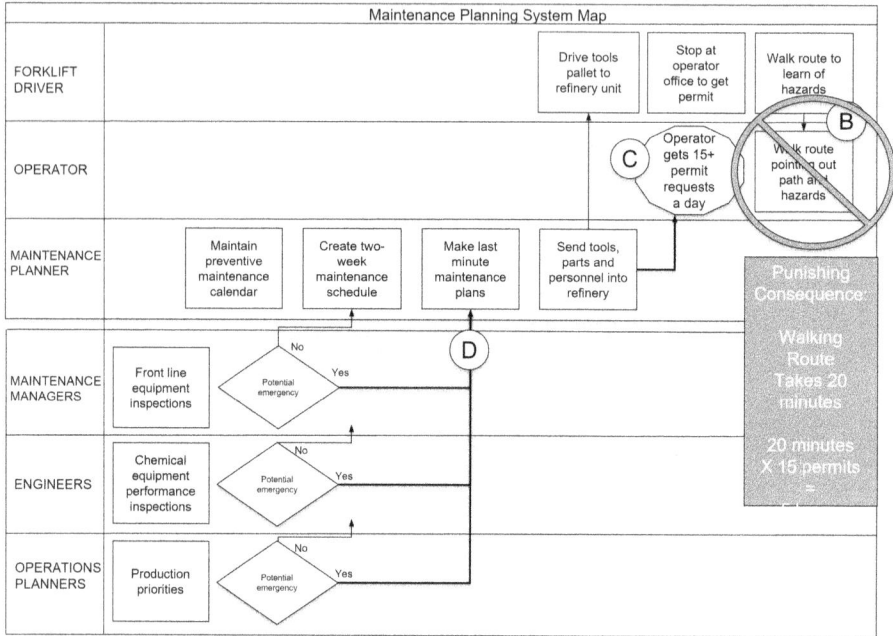

Figure 6.5   Map of Maintenance Planning System

punishing effect of the walkabout. Walking the route property took 20 minutes; multiply that times 15 permits. Operators were being asked to spend an average of five hours a day doing permits while trying to do their primary (full-time!) job of operating the unit. Were they doing the walkabouts? No. Would you? No.

At this point, the team continued their behavioral systems analysis (Ludwig, 2015), with a map of the greater system leading to warehouse deliveries for maintenance work and permitting. This started with the process of maintenance planning. These plans set into motion the delivery of construction tools within the refinery based on the maintenance to be conducted. Flowing downstream, this put warehouse delivery drivers in operator shelters asking for a permit to drive tools to where the maintenance was to be done.

So, let's check out the maintenance planning process. Maintenance planners kept calendars of the preventive maintenance required for the safe operation of equipment in a unit. This led them to create a two-week maintenance schedule for maintenance workers and sequence the delivery of the necessary tools and equipment from the warehouse. The schedule was shared with operators to anticipate permit requests on a reasonable controlled schedule. That is how the system was supposed to work. However, that's not how it *really* worked.

In addition to scheduling routine preventive maintenance, other maintenance personnel were doing equipment inspections for safety and reliability and finding unanticipated problems that needed mitigation. Those which did not constitute emergencies could be planned into the schedule. However, equipment issues that could result in a safety or reliability problem were often prioritized to be done within a week, due to potential upsets to product or safety. These were added to the planned work at the last minute (see Circle D in Figure 6.5). On top of that, other functions also demanded maintenance work be done as a priority because of their immediate needs. Engineers inspecting chemical specifications found that their equipment needed calibration or adjustments; operations managers often called in immediate fixes needed to optimize product flow; and the warehouse itself sometimes tried to make up for backlog by delivering more items than scheduled. Construction projects added another level of activity as they delivered scaffolding, tools/equipment, and contractors to the refinery.

All of these priority demands required maintenance planners to add more work to daily schedules. All of these interlocks set the occasion for the maintenance planner to deliver, on average, 15 people a day to the front-line operators requesting a permit. The ultimate cause of the forklift driver nearly hitting overhead chemical piping was a planning system out of control (Deming, 1982, 1986).

7
BEHAVIOR CHANGE INTERVENTIONS

The trended behavioral data showed that workers were taking risks with their body positioning, especially around the pinpoint "Overextending when opening valves." This was a problem because many of the older valves needed "some convincing" to be moved. This meant that a lot of torque had to be used. If you overextend when trying to exert torque, that's a good way to get a strain in your shoulder extending into your lower back. Recent injuries proved this, with one even resulting in a fall and a nasty bump on the head. The team got together with management to decide what to do to reduce overextending. They took the easy route, but one they thought would be effective. "Let's make a big deal about it—have it come from management and people will do it," was the call. So, the plant manager made a big announcement and supervisors added overextending discussions to their toolbox talks. Follow-up data showed that overextending reduced and safe body positioning took over. Success!... That is, until management moved on to other priorities and the communications stopped. Overextending showed back up in the data. It didn't work. They needed to find a more sustainable solution.

The goal of all these analyses is to determine problematic sources of variance that account for trends in behavior and then target the real reasons for the at-risk behavior through intervention and evaluation. The singular purpose of behavioral safety is to increase safe behavior and decrease at-risk behavior— not just in the short term with just a couple workers, but in a way that sustains direct-acting contingencies on workers' behavior. To do this, we must consider tactics that target worker safety skills, situational awareness, and culturally relevant behaviors (e.g., peer-to-peer coaching), before moving on to changes in the workers' environment and organizational systems.

TARGETING REPERTOIRES OF BEHAVIOR

The most common tactics for behavior change in industry are antecedent interventions. These include variations of safety announcements in meetings, postings, and "awareness" sessions. These are cheap and can show short-term results. SENCO's BEES program (Behaviors Encouraging Employee Safety; CCBS, 2015b) often conducts "blitzes" in which the entire workforce focus on

DOI: 10.4324/9781003290711-7

one safety pinpoint for a certain period (e.g., "hand safety month"). Blitzes include morning "toolbox talks," where the day's tasks are reviewed and safety pinpoints are discussed; and signs at task locations reminding workers of the safety precautions around hazardous machinery. Indeed, in a review of safety interventions highlighted in CCBS accreditation applications (n = 51) from top behavioral safety programs, 98% of them described workforce-wide antecedent initiatives.

Geller and Ludwig (Geller et al., 1990; Geller & Ludwig, 1991) proposed a multiple interventions level (MIL) model (see Figure 7.1) to describe the predicted impact of large-scale behavior change initiatives. Interventions were divided into multiple tiers, each defined by effectiveness, intensity of personal contact, and cost per individual. Large-scale interventions (Level 1 in Figure 7.1), where the agent-to-target ratio is high (e.g., one safety

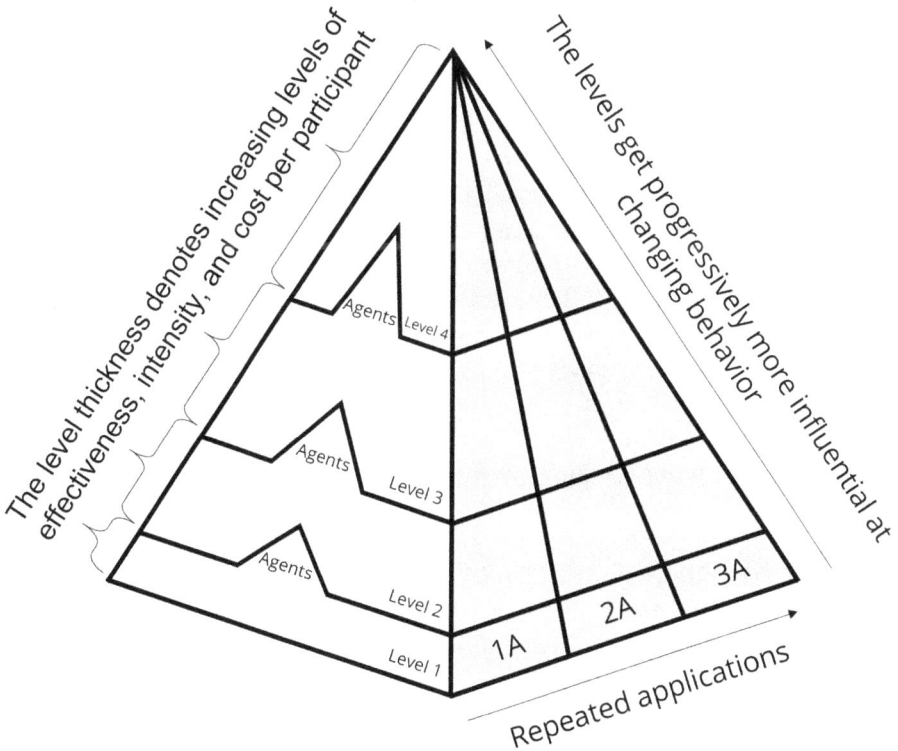

Figure 7.1   MIL Model

Adapted from "A Behavior Change Taxonomy for Improving Road Safety" by E.S. Geller and T.D. Ludwig, in M.J. Koornstra & J. Christensen (eds.). *Enforcement and Rewarding: Strategies and Effects*, pp. 41–45. Proceedings of the Organization for Economic Co-operation and Development International Road Safety Conference, Copenhagen, Denmark.

professional targeting 400 workers in a plant), typically are the cheapest per capita approaches and target the maximum number of workers. Examples of these are the corporate announcements and manager pronouncements common in the antecedent arsenal. These large-scale interventions can be effective at influencing individuals whose behavior has already been shaped. Got that? These Level 1 antecedent interventions only impact those who already know how to do the safe behavior and perhaps may influence them to practice what they know. Repeated applications of these interventions are necessary to combat the inevitable drift (Hyten & Ludwig, 2017) associated with antecedent-only interventions.

However, these large-scale antecedent interventions do not impact everyone. Indeed, a substantial number of individuals will not change their behavior in the face of big awareness marketing. These individuals require more individualized and costly interventions at Level 2. In industry, these take the form of crew-level meetings such as toolbox talks, where the agent-to-target ratio is much lower. These more intensive interventions can influence the behavior of even more individuals, but at greater cost. Daily toolbox talks absorbing thousands of labor hours can cost a company millions of dollars annually.

More intensive and expensive interventions at Level 3 are required when these group reminder processes fail to influence the behavior of the few workers who are "holdouts." There are many reasons why these individuals seem like "holdouts." They may not have the repertoire to work safely (e.g., new or transferred workers). If they have the repertoire, it is likely that their safe behavior has been punished; or alternatively, that their at-risk behavior has been reinforced by workplace contingencies.

When workers' safety repertoire has yet to be shaped, training is needed. Most training, unfortunately, takes the form of antecedent-only events targeting a classroom of workers away from the job site. Good classroom training should deliver contingency-specifying statements (Mawhinney, 2001), where instructions specify the safe behavior in the context of a specific situation (antecedent) to avoid a specific type of injury (consequence). Consistent with the MIL Model, this type of training—targeting larger numbers of individuals with less expensive methods (e.g., PowerPoint) —may influence the behavior of some of the individuals who may have some repertoire built from past experience. So, this is probably still a Level 2 intervention. Regardless, many workers will fall through the cracks and continue working at-risk. A more intensive intervention is thus needed for this smaller population of workers.

Behavior analysis has a history of training-to-fluency[SM-7.1]. In this type of training, individuals are given ample opportunity to practice the targeted behaviors and come in frequent contact with feedback to shape their repertoires (Binder & Sweeney, 2002). It is not sufficient to tell a worker a set of safe behaviors in a classroom setting. Rather, the worker must be trained in a way that increases the probability that they will respond when a situation requires specific safe behaviors. Workers also must be ready to respond safely in novel situations.

Science Moment 7.1
Fluency

Binder (1996, 1999) defines "fluency" as both the degree of accuracy and the speed of a target response. To develop fluency, training consists of three stages:

- acquiring a target behavior;
- practicing for fluency and endurance; and
- shaping trained fluent components to more complex behavior (Binder & Sweeney, 2002).

Developing safe behaviors with fluency training is appropriate because safe behaviors often must be performed with precision and fluency. As an example, a lab chemist must work precisely (e.g., measurements, handling of flasks) and fluently (e.g., timeliness of mixing materials) with chemical compounds to minimize potential risk. Safe behaviors also must be performed for an extended duration and/or without the loss of attention. For example, musculoskeletal injuries of the back often occur after extended periods of incorrect lifting (BLS, 2019b). Therefore, fluency in proper lifting mechanics can mitigate potential injuries with the high frequency of lifts in warehouse settings.

More complex activities are also appropriate targets for fluency training. If you bring simple behaviors up to fluency first, you can start adding them up into more complex classes of behaviors that can then be shaped to mastery (Binder, 1996). For example, if an individual is trained to fluency in individual behaviors in the smaller components of a drill switchout, they may perform better when faced with a more complex full drilling procedure. Finally, some really good news: building fluency in simple behaviors increases the likelihood that they will be used in novel combinations and situations. An electrician who has been trained to fluency in working safely around and identifying *live* electrical currents in his plant could generalize that repertoire to novel situations.

Fluency training sees workers engage in simulations of common hazardous tasks guided by a trainer, who provides frequent feedback until the task safety behaviors are mastered across multiple trials. Fluency training would satisfy the MIL Model's (Level 3) prediction of a more effective behavior change intervention because of the individualized practice and feedback. However, the associated costs in labor hours and lost productivity are typically not acceptable in a business setting trying to scale this method across several types of tasks with hundreds of workers.

Many companies engage in on-the-job (OTJ) training instead. In OTJ training, the worker engages in the job task while being observed by a tenured expert colleague or supervisor, is given frequent feedback, and is then certified once they satisfy selected criteria of successful task performance. Unfortunately, many OTJ programs are not sufficiently designed to shape fluency because trainers are not equipped with the time and behavioral tools necessary to pinpoint, observe, correct, and reinforce sufficient exemplars in a robust shaping process (Binder, 1996). What is more likely is the trainer teaches their own approaches to tasks, not those directed by the operating procedures (if there are any), and thereby passes on some of their own at-risk behaviors.

Companies that recognize the need to shape employee safety behavior to a level of fluency invest in structured mentorship programs that equip mentors with the tools of behavioral coaching:

- a pinpointed list of behaviors with competency criteria that must be accomplished by the worker;
- sequenced performance levels requiring higher and higher levels of mastery to be certified by the mentor;
- structured practice and feedback sessions with the worker spanning multiple weeks; and
- mentor training requiring their own mastery of these behavioral tools (i.e., behavioral observation, feedback skills, shaping, and evaluation; CCBS, 2015a).

DISCIPLINE PROGRAMS

When all else fails . . . If an employee repeatedly exhibits at-risk behavior around a high potential hazard, breaking what has come to be known as a cardinal rule, the most individually intensive, one-on-one intervention is required (Level 4). This typically is when the individual, following repeated violations of safety policy, is subject to the company's justice or discipline programs. Popular "three-strikes" programs (Guo et al., 2018) start with one-to-one "counseling sessions" with a direct supervisor who makes contingency-specifying statements alerting the worker of penalties designed to punish the at-risk behaviors. If the worker continues to engage in the at-risk behaviors (and gets caught), they can expect unpaid time off (strike two) or termination of employment (strike three).

Certainly, discipline programs are the most personally intensive types of behavior change interventions and can be costly in many ways. The cost to the worker is obvious; but the company must then bear the cost of hiring a new worker and training them to replace the lost productivity of the fired worker. This process also exacts a cost for the disciplining supervisor, who must have uncomfortable conversations, harm another person's livelihood and suffer repercussions from workers who may be upset about the firing of their colleague. All these costs quite often serve to punish actions on the part of supervisors to engage in discipline and may negatively reinforce behaviors such as "looking the other way" when a safety violation is discovered. Note that when these avoidance behaviors are shaped among the supervision of a company, the threat of discipline becomes a less probable and thus a less powerful contingency.

One additional consideration when discipline or, more specifically, the threat of discipline is the primary tool to influence the behavior of employees is that supervisors get shaped to use negative reinforcement as the primary method to manage safety (Ludwig, 2018). Supervisors seek out at-risk behavior and become quite fluent at scolding workers when caught. Workers then engage in the safe behavior to avoid the supervisor's scolding; their behavior has become negatively reinforced. Therefore, the supervisor becomes a discriminative stimulus, signaling that at-risk behavior will be associated with the punishing consequence of scolding. However, discipline and scolding are only effective when the disciplinarian is present. When supervisors are not present, while in meetings or doing paperwork, the discriminative stimulus is not present and workers will thus engage in behaviors consistent with the work environment. Workers master a quick change to safe behaviors when supervisors enter the workspace. Workers also master signals to alert each other that the supervisor is coming (Ludwig, 2018)!

Instead, a system to positively reinforce safe behaviors and correct at-risk behaviors in a culturally relevant and sustainable way is a more behaviorally sound approach. As noted, the most prominent way in which most behavioral safety programs successfully engage in behavior change is through a peer observation and feedback process. These peer interactions alone have been shown to impact the behaviors of both the observed and the person conducting the observation (Sasson & Austin, 2005).

PEER OBSERVATION AND FEEDBACK

Peer feedback delivers probable, prompt, and personal consequences to workers in the context of the very work environment in which at-risk behaviors are maintained. Think about that for a moment: it's quite powerful! When the observer who is giving feedback acknowledges safe behaviors, the worker experiences direct reinforcement for these actions. When the workforce engages in a high level of participation in a behavioral safety process, the frequency of reinforcement can be a powerful behavior change tool (Geller, 1996, 2005b; McSween, 1995, 2003).

Similarly, when an at-risk behavior is addressed with corrective feedback (Geller, 1996, 2005b), the worker is provided with contingency-specifying statements (Agnew & Redmon, 1993; Blakely & Schlinger, 1987; Mawhinney, 2001; Schlinger & Blakely, 1987), noting relationships between antecedents, at-risk behaviors, and consequences such as injury and discipline. The safe behavior is then discussed with the expectation that the worker will agree to try this alternative behavior in the future. Corrective feedback, however, may be ineffective if it is not followed by reinforcement when the worker ends up doing the safe behavior. Without reinforcement, the worker's behavior will drift back to the at-risk behavior.

Now get ready for a barn-burner: research suggests that the biggest impact of the peer observation and feedback process may not necessarily be on the recipient of the feedback. We've found that in fact, the biggest behavior change happens with the person conducting the observation and giving the feedback. Yup, the observer is the person who benefits most. Research on the "observer effect" suggests that the observer is three times more likely to change and maintain their behavior than the recipient of the feedback (Sasson & Austin, 2005; also see the literature on observational learning: Greer et al., 2006; Townley-Cochran et al., 2015).

When watching another worker doing a task, an observer may see examples of safe behavior being modeled by their peer. We know modeling changes behavior (Baer et al., 1967). They may also witness at-risk behaviors, some of which they may do themselves. However, because they are not caught up in the flow of doing their own physical tasks, they can observe these behaviors in the context of the other worker's environment (which usually resembles their own). The observer thus experiences both the immediate context (e.g., the task, physical environment, tools, hazards) and more distal antecedents (e.g., training, instructions, supervision, past experiences) of the behavior. Because they are not focused on doing the task as they would in their own job, they are able to see the behavior as it interacts with the hazards of the workplace. From this perspective, they may better realize the severity of potential injury as a possible consequence of the behavior being observed (or the avoidance thereof).

When giving feedback, the observer can use the context they experienced to discuss why the behavior is happening, perhaps even writing this down as a comment. They can also discuss what could be done to address these situations. Having given the feedback, the observer, after returning to their work, will be more likely to notice the situations surrounding their own tasks and how these can lead them to risk. Thus, they are more likely to do something about these problems. Further, when an observer coaches a peer on at-risk behavior, they probably won't go back to work and engage in that at-risk behavior themselves. They don't want to look like a hypocrite to others (or themselves—a concept known as "cognitive dissonance").

BEHAVIORAL ENGINEERING

Okay: now we've addressed all the cheap and (sometimes) easy interventions that impact individual worker behavior. In each of these, we are targeting individual workers with information, training, discipline, or feedback. Next we turn our attention to the types of interventions that impact all workers, present and future, who will do the job. Our goal is to drive sustainable behavior change by adapting the very environment and systems that put the worker in a position to take the risk.

Direct observation of worker behavior produces trends of at-risk behavior over time that provide a wealth of information worthy of analysis. These observations also provide insights about the direct environment workers behave in. Functional analysis tools—behavioral and otherwise—provide insights into the behavior of other people in other jobs whose interlocking decisions and actions might put the frontline worker in a position to take a risk. This gives us visibility into the systems and processes that create the environment in which workers work and take risks. The objective now is to adapt these systemic processes of interlocking behavioral contingencies to remove the need to take a risk (Step 5; Figure 1.4 Evergreen Model).

First, a word of caution. A common problem with safety interventions is that managers overengineer their solutions, often requiring more effort from workers by implementing more rules, safety processes, protective clothing, guards, and equipment. These cumbersome requirements increase response effort (which our science calls "response cost") when workers engage in safe behaviors within these additional protocols. The extra response cost can punish these new behaviors, which leads workers to drift back to their old at-risk behavior and, most likely, pencil whip the processes. The company then puts additional controls in place to compel the behaviors through more audits and supervision, along with sanctions and discipline.

These additional rules and procedures require a negative reinforcement scheme that is costly and difficult to enforce. The threat of punishment by supervisors (who are not around) is not powerful enough to overcome the probable, prompt, and personal punishment directly related to the increased response cost of engaging in these new requirements. To make things worse, workers are less likely to report at-risk behaviors and minor incidents for fear that if they do so, management will demand even more cumbersome safety processes. Designing additional safety processes may be warranted; however, it is strongly recommended that new processes be designed with sufficient consultation from workers, who can provide input on the extra response cost required.

The National Institute of Occupational Safety and Health (NIOSH) defines intervention priorities in its publications listing a "hierarchy of controls" (NIOSH, n.d.). Administrative controls (rules and procedures) and personal protective equipment (PPE) fall at the bottom of their hierarchy. Instead, NIOSH's "Prevention through Design Initiative" (NIOSH, 2011) encourages efforts to anticipate and design out hazards to workers through adaptations to work methods and operations, processes, equipment, tools, products, technologies, and the organization of work. If you can't eliminate the hazards, the most effective approach is to engineer the need for human behavior out of tasks that involve contact with serious hazards. If this is not possible, equipment and facilities should be adapted—including, for example, installing equipment guards, making ergonomic improvements, replacing failing equipment, and fixing hazards around the facilities. If hazards cannot be removed, then engineering of equipment, tools, and workspaces should be prioritized to isolate workers from the hazards (NIOSH, 2011). These changes are counter to the overengineering problem, as changes here should reduce response effort, response cost, and hazard exposure.

While these engineering approaches to safety may not seem "behavioral," they require a multitude of interlocking behaviors, starting with an effective system of documentation and communication from the frontline through management. Organizational functions such as engineering, maintenance, and procurement must then prioritize, design, procure, build, and deliver these solutions. The projected costs, disruptions, and timelines of engineered solutions that eliminate, substitute, and/or remove workers from hazards often serve as discriminative stimuli signaling punishment to decision maker behaviors.

Gilbert's Behavior Engineering Model (1978) offers an instructive framework for considering and prioritizing behavior change interventions targeting the opportunities discovered in the behavioral observation and functional analysis processes. In his model, he distinguishes between engaging environmental supports of behavior and the person's repertoire of behavior. While both offer behavior change potential, Gilbert is clear that environmental supports offer more impactful and sustainable interventions that more broadly impact the workforce over time than the more temporary person-specific interventions aimed at impacting the behavioral repertoires of workers through antecedents (e.g., training) and hiring.

A behavioral engineering approach to safety seeks to design the direct-acting contingencies present in work environments that put the worker in a position to take the risk. We can too often rely on the actions of direct supervisors or fellow workers to prompt safe behaviors and/or provide consequences for at-risk behavior. However, as we've noted many times, these coaching behaviors drift over time; while production demands and cost cutting differentially reinforce other, more at-risk behaviors. Therefore, simple reliance on worker/supervisor prompting does not produce sustainable change.

Here is the important point to get your head around: it is your system that creates the environments for the worker—the environments that the worker works and takes risks in. If you change the environment, you can change the behavior. You can sustainably change the environment by changing your systems. The interlocking systems that establish and maintain the contingencies governing worker behaviors must be part of any meaningful intervention. Behaviors of people far away from the front line, within the support functions of an organization, intentionally or unintentionally change the direct-acting contingencies of the workplace and have a greater impact on a greater number of workers than you may realize.

Changing both the direct-acting contingencies in the immediate environment of the worker and the systemic interlocking behavioral contingencies of the larger organization should be part of any intervention package. Table 7.1 offers some common examples of the linkages between the direct-acting contingencies and systemic interlocking behavioral contingencies that together serve to maintain the safe behavior of frontline workers.

Don't forget to include workers in your intervention design! These interventions should be a product of the behavioral systems analysis conducted after trending behavioral data or when planning upcoming tasks. Therefore, the group process that produces the Antecedent-Behavior-Consequences (ABC) and behavioral systems analysis should include knowledgeable representatives from organizational functions that interlock with frontline worker contingencies (Blasingame et al., 2014). While professionals from these functions should "own" the system design and implementation that results from this analysis, action plans must also include workers in the design. As argued, frontline workers are the only ones who experience the direct-acting contingencies that maintain or fail to reinforce safe behaviors. They can predict if fellow workers will participate in the new design (or pencil whip it) and help determine if the design will reliably result in safe behaviors.

It is important to note that we cannot expect an employee team to take responsibility for the implementation of all levels of behavioral interventions. Indeed, most systemic improvements require the professionals and managers throughout the organization to make decisions and take action. When solutions are identified during ABC and behavior systems analysis, the team should apply a classification to denote which "level" of organizational actors are responsible for implementation of the contingency improvement:

- Level 1: Frontline employees and/or employee teams are empowered to study the issue, procure materials (if required), and make changes. These can and should happen right away.
- Level 2: Safety managers and operations supervision are tasked with making the changes happen. These may take some additional time to study, gain approval and implement.
- Level 3: The changes require system processes that may involve multiple functions (e.g., maintenance, engineering, procurement) across the site or company. Because of this, top leadership and cross-functional teams are typically engaged with employees to study and improve targeted behavioral systems. Because budgetary and other downstream impacts must be considered, these changes typically take a long time to accomplish.

Table 7.1 Linkages Between Direct-Acting Contingencies and Systemic Interlocking Behavioral Contingencies

Safe behavior (response class listed)	Direct-acting contingencies	Interlocking (system) contingencies
Pace of work	Providing adequate time to do the job reduces negative reinforcement of hurried work behaviors. Allow workers variance for work schedule disruptions if reported.	Engineers using time and motion data to design job tasks. Contracts written without incentives for above-average timely work. Manager incentives for production based on designed time and motion.
Ask for help completing a hazardous task Ask for a "watch" person when in a confined space or similar	Opportunity for behavior because enough workers have been staffed for the hazardous task. Removes punisher for requesting help from others who may be busy doing something else.	Manager/leader budgets for adequate staffing based on upcoming work with buffer based on data from past work upsets. Human resources forecasts absenteeism and turnover, and staffs accordingly.
Use proper tools, in working condition, for the job task	Provide and verify proper tools near the job task. Allow time at the beginning of the shift for workers to review task lists and retrieve proper tools.	Procurement inspection program finds and replaces broken and missing tools. Supervisors provide morning task lists and specify proper tools. Tool cribs are built near common tasks.
Follow operating procedure processes	Operating procedures do not punish safe behaviors with unreasonable response cost.	Engineers and managers engage with workers to evaluate and revise operating procedures.

Wear appropriate PPE for the hazards present on the task	Better-fitting, comfortable and functional (not interfering with the task) PPE, so not to punish its use.	Procurement and safety managers engage with workers to evaluate and adopt functional PPE.
Stop work when there are concerns that it presents too much risk	Clear stop-work processes and policies that do not punish worker or supervision.	Stop-work process documentation visibly approved by top executive. Human resources contact available for reports of incidents where stop-work requests were denied or punished.
Engage in permitting processes that share knowledge of risk, check tools and equipment, and adequate staffing	Adequate time designed into task planning for all involved in permitting.	Task planning process designed to allow for unplanned variance. Additional permitters available when the load requires.

Change at Level 3 can be complicated and costly. Also, change at Level 1 should be impactful and frequent. Therefore, the workforce can influence Level 3 initiatives by accomplishing many multiples of Level 1 changes delivering value to operations and potentially reinforcing leadership action. We invite you to go back to Table 6.2 showing the solutions the warehouse team came up with in their ABC analysis, and Table 7.1 above, and apply the appropriate "level" to each of these interventions.

8 EVALUATION

The only way to understand behavior is to change it. Murray Sidman (1960) wrote: "When an organism's behavior can repeatedly be manipulated in a quantitatively consistent fashion, the phenomenon in question is a real one and the experimenter has relevant variables well under control" (p. 85). Indeed, all the principles, practices, and citations in this book are for naught if critical safety behaviors are not shaped, reinforced, and verified. Thus, in order to be sure we have mastered our understanding of safety behaviors, we must engage in an evaluation of behavior change as well as the impact of these new safe behaviors on injuries.

Evaluation is critical to confirm that all the steps in our behavioral safety processes did, in fact, change behavior (Step 7; Figure 1.4 Evergreen Model). We do not "train and hope" (Stokes & Baer, 1977; Stokes & Osnes, 1989)—referring to the all-too common practice of finding at-risk behaviors, then throwing a training intervention at them and considering the problem solved. Instead, behavioral best practices based on science require evaluating both if meaningful change occurred in critical pinpointed behaviors and the resulting impact on injuries (Gilbert, 1978).

Strict evaluation is one of the primary standards adopted by the Cambridge Center for Behavioral Studies (CCBS) (behavior.org) Commission on Behavioral Safety. The CCBS Standards (CCBS, 2022a) generally require that a behavioral safety program be:

- based on the science of behavior analysis;
- engaged with integrity (i.e., you are doing what you say you are doing); and
- associated with a reduction in injuries.

The Commission's mission is to disseminate these evidence-based best practices to others seeking to successfully implement behavioral safety principles. While we cite the science of behavior analysis, more specific best practices came from our learnings from CCBS accredited behavioral safety programs. Therefore, we conclude this book with a discussion of some of the best-in-practice programs evaluating behavior change and confirming injury reduction.

DOI: 10.4324/9781003290711-8

INTEGRITY

When a process with integrity runs as intended, people are doing what is requested with mastery and your results are replicable. All too often, we skip this important assessment and assume that what we put in place is being done (a big assumption), and that it is being done right (a bigger assumption). If we don't measure integrity, it probably means our process doesn't have it. Measures that assess the quality of execution within behavioral safety programs are also referred to as "process measures" (Geller, 2001b, 2002a, 2002b), as they confirm critical steps in the process—observations, feedback, shaping, analysis, and interventions (a.k.a. "follow-on actions"—are conducted as planned.

PARTICIPATION

Participation and contact rates are process measures that seek to assess the quantity of observations, often as a percentage of the workforce. Counts of observations tracked monthly can provide information on trends in participation. When a behavioral safety program is first established, the first few months of observation counts may be quite low as workers have yet to sample the process, gain some mastery in peer observation and feedback, and see their participation as helpful toward uncovering and solving problems that lead to risk. However, the second and third months will likely show higher and higher levels of participation (often doubling or better), as individuals experience the process as relatively benign, hassle-free, and worth attempting.

Most programs measure an upward trend in participation counts for six months up to two years, when a natural plateau occurs. This plateau represents the best capability of your behavioral safety process to engage a certain segment of the worker population. Yep, all the training, promotions, management support, blood, sweat, and tears have got you to right here. But doing more of those same things will not get you any further.

SDR Coating Company's accredited program (CCBS, 2015c) demonstrated a similar pattern of participation in its first three years (Figure 8.1). It topped out at around 40% participation, which is the best the program as it was practiced could achieve. And, by the way, 40% of the workforce doing an observation and giving feedback in a month is pretty darn good! As a benchmark, it is a goal for your program.

Best-in-practice behavioral safety programs tend to achieve 40% to 60% participation in their voluntary observation programs. However, recent evidence suggests that participation rates of between 4% and 8% can be more effective at reducing injuries than higher participation rates if observers are well trained and observations and feedback occur with quality and integrity (Spigener et al., 2022). There are numerous strategies to achieve quality levels of participation. Eastman Chemical's Acetate Fibers Division (AFD) steering committee

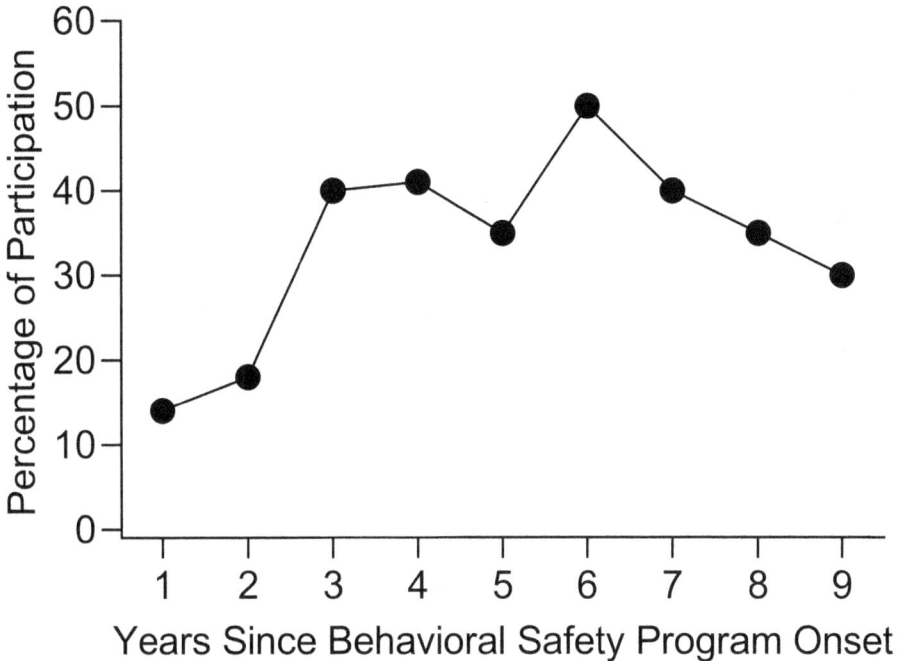

Figure 8.1   SDR Coating Participation Through First Nine Years

Percentage of participation on the y-axis and years since behavioral safety program onset on the x-axis. Data points indicate the percentage of participation per year. Data were adapted with permission from the CCBS. From "Behavioral Safety: An Efficacious Application of Applied Behavior Analysis to Reduce Human Suffering," by T.D. Ludwig and M.M. Laske, *Journal of Organizational Behavior Management* (Taylor & Francis, 2022).

decided to increase participation through a coaching process to improve the quality of observations. Frontline managers coached each employee during an observation, with the goal of coaching 100% of the AFD's approximately 600 operators/mechanics (CCBS, 2006). This coaching process allowed employees who had never done an observation to test out in the voluntary observation process at least once. The result of this observation coaching was an 18% increase in observations, from 2,917 in 2005 to 3,453 in 2006 (CCBS, 2015a).

Once participation rates have settled into a predictable trend, behavioral safety programs often measure "contact rate" as a percentage of monthly observations divided by the number of workers (e.g., Marathon St. Paul Park Refinery, Marathon Michigan Refining Division, Costain, Marathon Illinois Refining Division, Marathon Michigan Texas Refining Division: CCBS, 2022b). A rate of 1.0 indicates the month's observations are equal to the number of employees in the workforce. Programs typically set goals of achieving a contact rate of 1.0, suggesting that every employee had an "opportunity" to be observed that month.

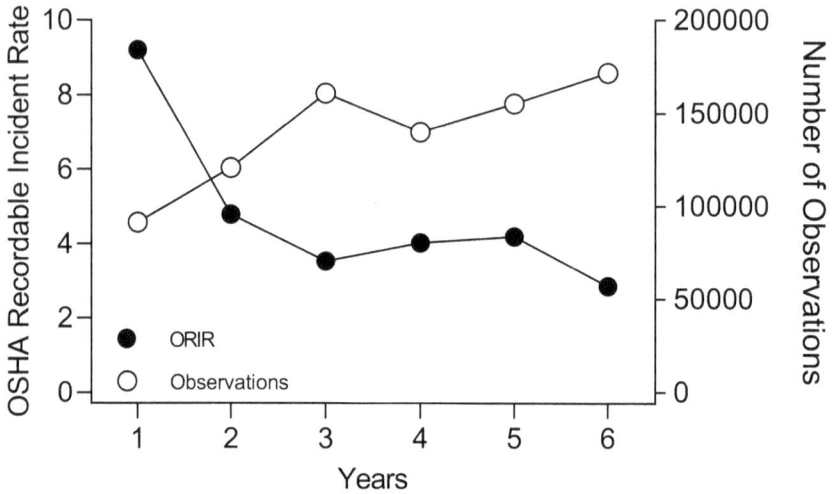

Figure 8.2 Supervalu Occupational Safety and Health Administration Recordable Rate and CAM Observations

OSHA recordable incident rate (ORIR) on the primary y-axis; number of observations on the secondary y-axis; and consecutive years on the x-axis. Black circles indicate ORIR. White circles indicate the number of observations. Observations data were adapted with permission from the CCBS. From "Behavioral Safety: An Efficacious Application of Applied Behavior Analysis to Reduce Human Suffering," by T.D. Ludwig and M.M. Laske, 2022, *Journal of Organizational Behavior Management* (Taylor & Francis, 2022).

There is a convincing relationship between observation counts and reduction in injuries. The empirical relationship between observation counts and injury rates can be seen by reviewing data from numerous CCBS accredited programs. Supervalu's Critical Actions Management (CAM) program (CCBS, 2013) demonstrates that increases in observations were associated with decreases in injuries over many years (see Figure 8.2).

The same relationship between observations and injuries is strongly demonstrated by Marathon Illinois Refinery Division's Forever Uniting Employees Lives Through Safety (FUELS) program (CCBS, 2019). Figure 8.3 compares the plant's ORIR with total annual counts of observations. Before implementation of FUELS, ORIR were high and variable (albeit decreasing). After the formation and implementation of FUELS, ORIR decreased as observations increased. Six years into program implementation, FUELS applied for and received accreditation for its behavioral safety process. In the following years, FUELS began to include its contractor workforce in its behavioral safety program and observations increased dramatically. This resulted in continued and sustained reductions in ORIR. In recent years, it appears that a decrease in observations has correlated with increases in ORIR, further demonstrating the inverse relationship between the two.

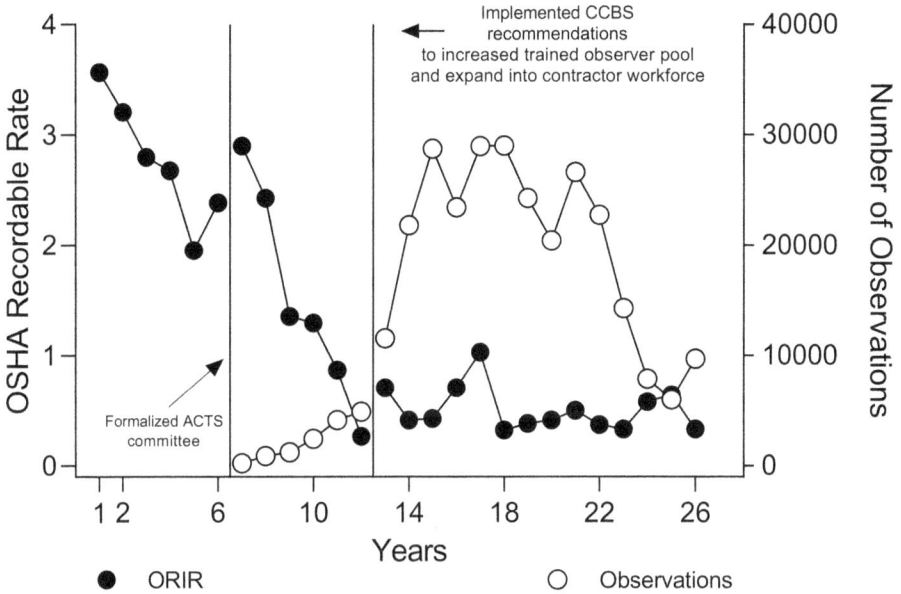

Figure 8.3   Marathon Illinois Refinery Division ORIR and FUELS Observations

ORIR on the primary y-axis; number of observations on the secondary y-axis; and consecutive years on the x-axis. Black circles indicate ORIR. White circles indicate the number of observations. The first phase line is where the Illinois Refinery Division formalized its Areas Communicating Trust in Safety (ACTS) committee. The second phase line indicates where the ACTS committee acted on the CCBS recommendations to increase trained observers and include contractors. Adapted with permission from the CCBS.

Quality Observations

A strict count of observations only shows the quantity of participation via submissions of observation data; it says nothing about the quality of that data. However, quality observations are more likely to accurately identify risks, provide information for more valid analysis and ultimately lead to successful behavior change interventions. Programs that produce low-quality observation data often show that nearly all targeted pinpoints are near 100% safe month after month. This near 100% safe certainly does not reflect the reality in the workplace. Low-quality data like this may suggest the workforce has not mastered the observation skills to get beyond a leniency bias (Saal et al., 1980) or concerns about documenting something "negative." Worse still, such invalid results may be the function of pencil whipping (Ludwig, 2014), which calls into question the overall integrity of your observation process.

For example, a contractor at a refinery promoted observations through a "Texas Hold-em" lottery, where workers earned a playing card for every observation turned in. At the end of the month, the worker with the best poker hand

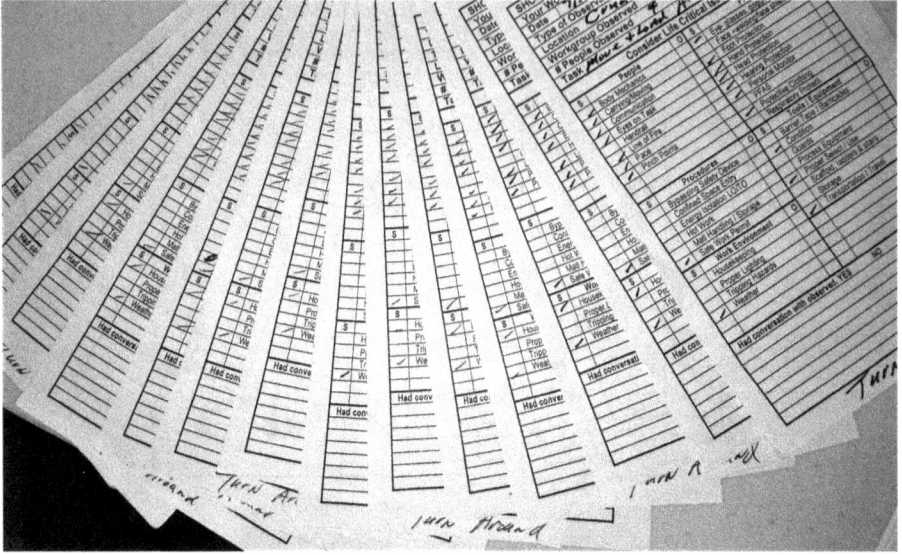

Figure 8.4   Pencil Whipped Observations

Photo credit: Timothy Ludwig.

won a prize. This resulted in a "scalloped" participation trend, with very few observations done at the beginning of a month and over 90% completed in the final week (many on the last day). Skinner also saw this type of response in rats and pigeons when putting behavior on a fixed interval schedule of reinforcement (refer back to Science Moment 5.4), similar to a quota being due at the end of the month. An examination revealed that large sets of observation cards were submitted by a single person at the same time, all with the same data—typically 100% safe (see a picture of these observation cards in Figure 8.4). Such response patterns are typical of incentivized participation campaigns and result in low-quality observations. To combat pencil whipping, many programs do random audits of observation cards to find response patterns that suggest many cards were marked at the same time and submitted together.

Programs should adopt integrity measures to assess observations for quality. One method to measure quality is to count the percentage of submitted cards that contain actionable comments (typically a comment section is included in the card; see Figure 8.5). These comments may document the risk in more detail (e.g., "Worker flipped up face shield when it got foggy, said he couldn't see"); describe the "barrier" that promoted the risk (e.g., "Pulling fuel hose upstairs is only way to fuel tank"); or suggest an on-the-spot solution discussed between peers (e.g., "Picked up can of bolts, approximately 65lbs. Told him to put on pallet and use telehandler or use another person to help carry"). Examples of safe behaviors can also be written in the comments (e.g.,

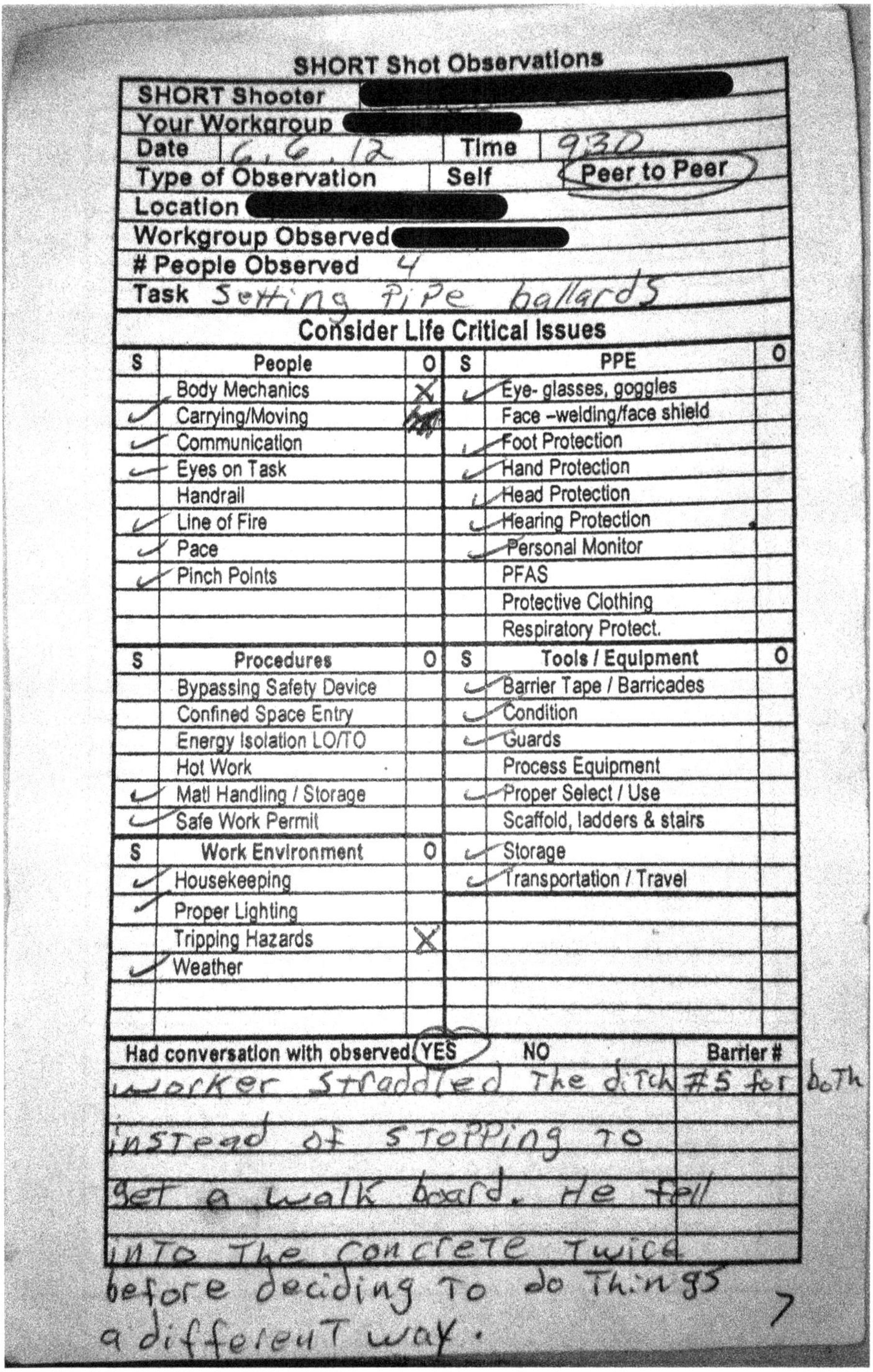

SHORT Shot Observations

SHORT Shooter	████████████
Your Workgroup	████████
Date	6, 6, 12	Time	930
Type of Observation	Self	(Peer to Peer)
Location	████████████
Workgroup Observed	████████████
# People Observed	4
Task	Setting Pipe ballards

Consider Life Critical Issues

S	People	O	S	PPE	O
	Body Mechanics	X	✓	Eye- glasses, goggles	
✓	Carrying/Moving	✗		Face –welding/face shield	
✓	Communication		✓	Foot Protection	
✓	Eyes on Task		✓	Hand Protection	
	Handrail		✓	Head Protection	
✓	Line of Fire		✓	Hearing Protection	
✓	Pace		✓	Personal Monitor	
✓	Pinch Points			PFAS	
				Protective Clothing	
				Respiratory Protect.	
S	**Procedures**	**O**	**S**	**Tools / Equipment**	**O**
	Bypassing Safety Device		✓	Barrier Tape / Barricades	
	Confined Space Entry		✓	Condition	
	Energy Isolation LO/TO		✓	Guards	
	Hot Work			Process Equipment	
✓	Matl Handling / Storage		✓	Proper Select / Use	
✓	Safe Work Permit			Scaffold, ladders & stairs	
S	**Work Environment**	**O**	✓	Storage	
✓	Housekeeping		✓	Transportation / Travel	
✓	Proper Lighting				
	Tripping Hazards	X			
✓	Weather				

Had conversation with observed	(YES)	NO		Barrier #

Worker straddled the ditch #5 for both
instead of stopping to
get a walk board. He fell
into the concrete twice
before deciding to do things
a different way.

Figure 8.5 Completed Checklist with Comments

Photo credit: Timothy Ludwig.

"After high readings on benzene meter, the job was stopped until proper PPE was determined"; CCBS, 2012c, p. 41).

Recent research conducted in our behavioral safety lab at Appalachian State University looked at three years of behavioral observation data from a large business division of a chemical manufacturing company to see if an observation was related to a reduction in the probability of an injury or near miss (Matthews, 2022). We found that observations did not predict decreases in the probability of injuries. However, when we only looked at quality observations, the story changed. Observations with comments were found to predict a reduction in the probability of an injury over the next five days. This was especially true when the comments were longer in length.

Another measure of quality is the percentage of observations that identified at-risk behavior. As noted earlier, a good pinpoint is one that finds risk when added to an observation card. A quality observation will find, document, and provide feedback on at-risk behaviors. Over a 15-year span, Eastman Chemical's AFD tracked the number of at-risk behaviors noted by observers on their observation cards. In the first two years, the data suggest that workers avoided reporting at-risk behaviors. However, as workers grew more trusting of the process and saw how injuries were being reduced, an approximately 600% increase in at-risk behavior identification was achieved by the third year, which nearly doubled again in the following years. The workforce of around 800 eventually reported nearly 1,500 at-risk behaviors a year. Figure 8.6

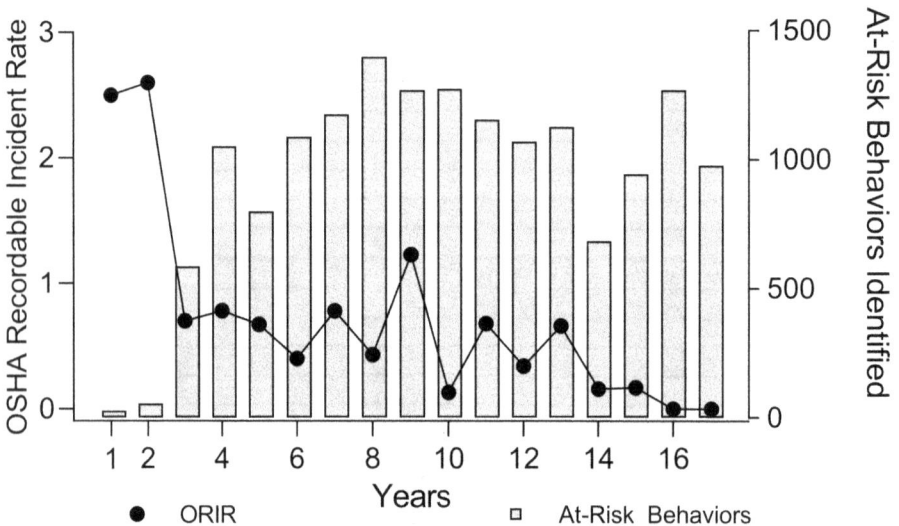

Figure 8.6   AFD At-Risk Behaviors Identified and ORIR

ORIR on the primary y-axis; at-risk behavior identified on the secondary y-axis; and consecutive years on the x-axis. Black circles indicate ORIR. Gray bars indicate the number of at-risk behaviors identified. Data are adapted with permission from the CCBS.

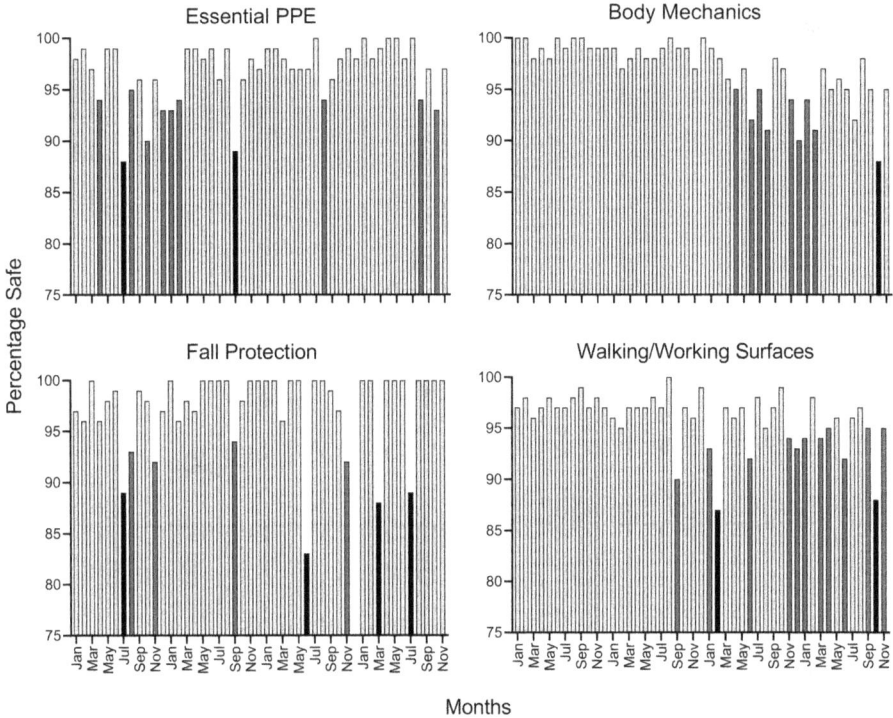

Figure 8.7   Aware Trending of Behavioral Pinpoints

Percentage safe behavior on the y-axis and consecutive months on the x-axis. Pinpointed observations below 95% safe are displayed in dark gray. Observations below 90% safe are displayed in black. Data were adapted with permission from the CCBS. From "Behavioral Safety: An Efficacious Application of Applied Behavior Analysis to Reduce Human Suffering," by T.D. Ludwig and M.M. Laske, 2022, *Journal of Organizational Behavior Management* (Taylor & Francis, 2022).

shows this trend overlaid with Eastman AFD's recordable injury rate (CCBS, 2015a), suggesting a relationship between finding risk and reduction in injury rates.

Figure 8.7 shows the monthly percentage-safe of observations conducted within St. Paul Park Refinery's All Work At-Risk Eliminated (AWARE) program (CCBS, 2020b). Monthly pinpointed observations below 95% safe are displayed in dark gray and observations below 90% appear in black, showing the identification of risk at least 10% of the time. These data suggest the AWARE observation process was successful at finding meaningful risk in 22 of the 36 (61%) months. They did a particularly good job at identifying risk during most months in the last two years of this data trend.

Marathon Michigan Refinery's Circle of Safety (COS) program utilizes a severity index to categorize the severity and probability of an incident based on an at-risk behavior (CCBS, 2020a). COS observers rate severity

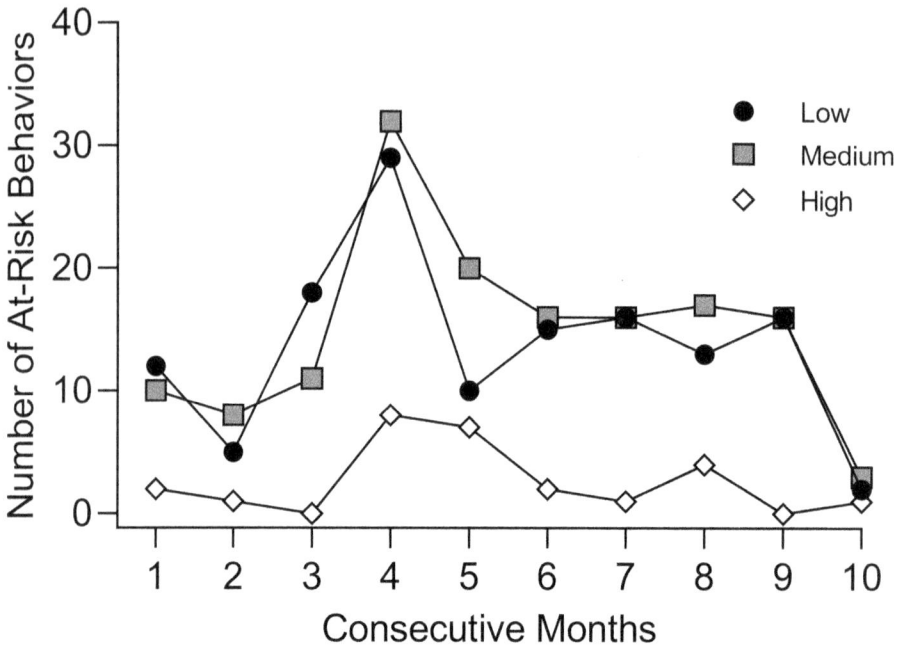

Figure 8.8   COS At-Risk Behavior by Potential Event Severity

Number of at-risk behavior on the y-axis and consecutive months on the x-axis. Black circles indicate at-risk behavior with low severity potential. Gray squares indicate at-risk behavior with medium severity potential. White diamonds indicate at-risk behavior with high severity potential. Data are adapted from the CCBS.

based on the task being completed, potential environmental hazards, and the likelihood of an incident. These severity ratings prompt further assessment of root causes through functional analysis. Figure 8.8 sets out COS at-risk behaviors trended by severity potential—a very good sign that they are getting quality observations.

A common method to increase the frequency of quality observations is through observation training and coaching. The St. Paul Park Refinery AWARE program, the SENCO Behaviors Encouraging Employee Safety (BEES) program, and many other programs (CCBS, 2005, 2017b, 2018b, 2018c, 2020b) coach observers to shape and verify observation skills. A coach will watch the observer do observations and complete a coaching checklist (see Figure 8.9). After the observer completes their observation and feedback session, the coach will review the coaching checklist, providing the observer feedback and suggestions. Coaching not only reinforces quality observations, but also is intended to shape quality interactions during the feedback portion of the session. As previously seen in Eastman Chemical's AFD, observation coaching also increases observation participation.

SENCO Facilitators Coaching Guide
Name of Coach_____ Date_____

In an effort to improve the overall quality of observations this guide was developed to help the various BBSCAP facilitators critique their trained observers. This in turn will:

- Help increase the quality of the observations being performed.
- Drive more communication and proper specific feedback.
- Positively motivate observers.
- More facilitator and observer interface
- One on one coaching.
- Show the facilitator what they need to improve on or stress in their training programs.
- Help improve the skills of the observers.

The coach should only take notes during the evaluation. Let the observer do the talking.

Introduction	Yes	No
1. Observer asked permission to do observation.		
2. Observer Explained the process.		
3. Observer explained the feedback process		
Feed Back		
4. Observer discussed specific safe behaviors first.		
5. Observer avoided using loaded words		
6. Observer discussed specific At-risks seen.		
7. Observer asked for commitment to working safe.		
8. Observer kept discussion positive		
9. Observer promoted discussion by asking questions.		
10. Observer checked only the parts of the form that applied to the task.		
11. Observer listened to answers, made sure employee understood.		
12. Observer explained Is follow up needed		
General		
13. Observer legibly filled out all applicable portions of the form		
14. Observer filled out comment section of form		
15. Observer treated worker like the worker wanted to be treated.(Platinum)		
16. Upon completion observer showed employee finished form		
Comment		

Note a QUALITY observation is;

- A Quality Observation is Clearly Written
- A Quality Observation has All Information Filled in
- A Quality Observation Includes Communication with the Person being Observed
- A Quality Observation that indicates an at Risk Situation, include a Barrier and Explain the Unsafe Act
- A Quality Observation is Turned in in a Timely Manner

Figure 8.9   SENCO BEES Observation Coaching Checklist

Figure is reprinted with permission from the CCBS.

Insightful Analysis and Follow-up Actions

When observations identify risk and provide adequate comments describing the risk and systemic barriers, we can better analyze why the at-risk behavior is occurring and then design better interventions that change behavior and reduce injuries. Quality observations lead to quality data that lead to higher-quality functional analysis of risk and thus more efficacious interventions that make a difference.

The frequency and quality of analysis (eg, Antecedent-Behavior-Consequences (ABC) analysis, behavioral systems analysis) can be tracked as a measure of integrity. Costain's behavioral safety program uses a team of behavior management experts (internal consultants) to train project teams to conduct ABC analysis at their construction sites. This team can even be called in to facilitate these ABC analysis sessions to mitigate risk in current or upcoming tasks (CCBS, 2018b). Therefore, ABC analysis is one of Costain's key performance indicators in its safety program—one it holds projects accountable for on its dashboards.

As ABC analysis reveals potential causes of at-risk behavior, follow-up actions are used to intervene on the risk. Completion of these actions can be yet another measure of the integrity of a behavioral safety program. In addition to tracking the completion of action items, Costain (CCBS, 2018b) publicizes these accomplishments to the workforce in the form of "You said, we did" postings at project sites (see Figure 8.10).

These measures of integrity (i.e., participation, quality observations, finding risk, ABC/behavioral systems analysis, follow-up actions) are not exhaustive. Evaluate the extent to which the various elements of behavioral safety programs are being conducted with frequency and quality. The reason we make the effort to do these process measures is to give our behavioral safety programs internal process feedback (Brethower & Dams, 1999) that facilitators and sponsors of such programs can use to improve their processes. With these measures, we can direct improvements when data

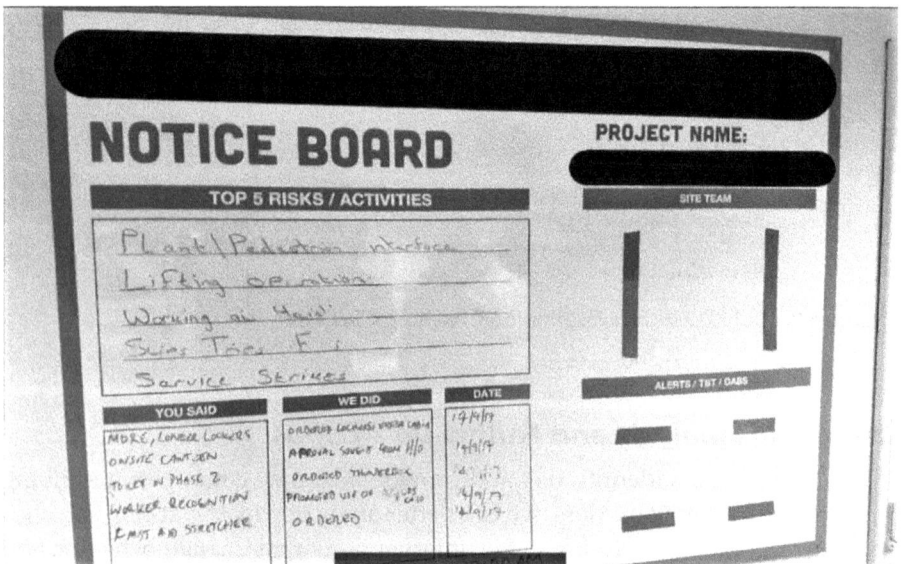

Figure 8.10   Costain "You Said, We Did" Action Completion Posting

Image is reprinted with permission from the CCBS.

suggests a drift in engagement and quality. However, our data also show that when the integrity of behavioral safety programs is strong, behaviors change and injuries are avoided.

BEHAVIOR CHANGE

The whole point of investing in behavioral safety is to produce sustained behavior change in the workforce to maintain safe work practices. To evaluate behavior change, we first need to establish a baseline to track where we came from. We establish a baseline by observing newly pinpointed behaviors and finding the current level of at-risk behaviors in the workforce. We can then compare the level of risk before and after our interventions to see if they had a desirable effect. If interventions do not successfully cause a change in targeted behaviors, further analysis and intervention design are required. Behavior change interventions can take many forms, as noted in earlier chapters. In this section, we provide evidence of behavior change in some of the most prevalent intervention strategies.

Behavior Change Due to Antecedents

Often antecedent-based interventions result in behavior change due to the extra attention on a pinpoint (e.g., training, prompts, awareness sessions, leadership messaging). However, continued observations reveal a drift back to risk after the antecedent blitz subsides. Therefore, these cases also require further analysis to design more sustainable systems interventions.

A good example of a program seeing change after antecedent blitzes only to experience drift is the AWARE program at St. Paul Park Refinery (CCBS, 2020b). Behavioral observations of its body mechanics pinpoint showed a downward trend over eight consecutive months (see Figure 8.11), eventually dropping below the percentage safe goal.

To address this risk area, AWARE engaged the workforce in training focused on body mechanics. The training resulted in the body mechanics percentage safe moving back above the goal. However, the following months showed a drift back to prior risk levels. It should also be noted that the training and new pinpoints may have eventually sensitized observers to ergonomic risks that they may not have previously perceived. Therefore, they may have started reporting at-risk behavior that may have existed in the past, but was not considered a risk. Regardless, the AWARE team saw they needed to continue intervening on body mechanics until the behavior change was sustained.

The AWARE team broke down the body mechanics pinpoint into more specific ergonomic movements and adapted their process to begin observations of these new pinpoints. As they trended the specific pinpoints, they found that "overextending" and "rushing" were the highest categories of

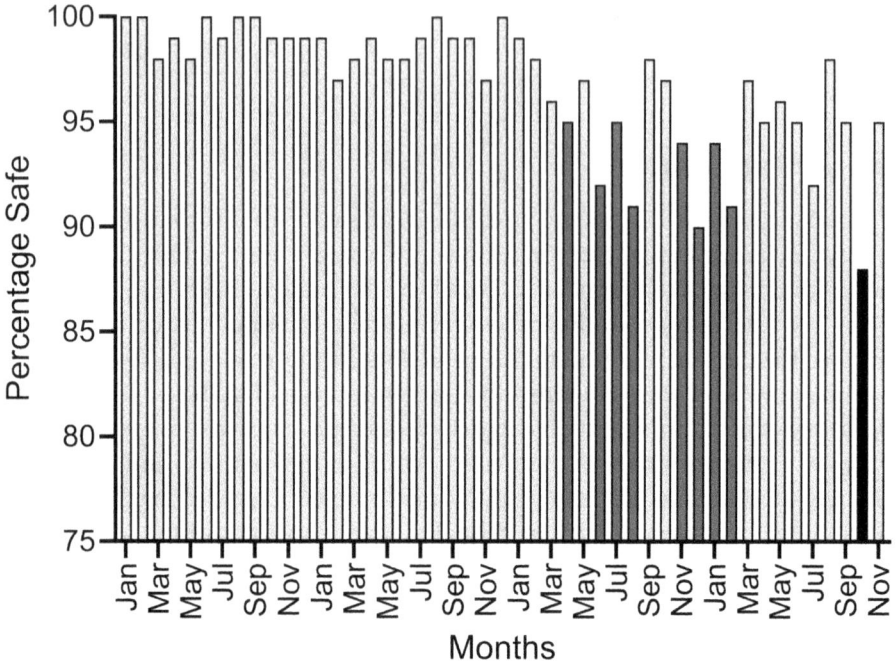

Figure 8.11   AWARE Body Mechanics Pinpoint

Percentage safe behavior on the y-axis and consecutive months on the x-axis. Pinpointed observations below 95% safe are displayed in dark gray. Observations below 90% safe are displayed in black. Data are adapted with permission from the CCBS.

body mechanics risk (see Figure 8.12). The analysis was shared with leadership, who agreed to make statements about rushing and overextending during pre-shift meetings. These manager antecedents resulted in several months in which observations of overextending showed a positive change up to 100% safe reaching. However, after several months of safe reaching observations, another period of drift was witnessed and overextending risk returned. This behavioral drift among the workforce was most likely related to behavioral drift among managers who, over time, moved on to different topics and stopped making statements regarding rushing and overextending. Without sustained manager antecedents, risk returned.

Eventually, the AWARE team conducted an ABC analysis and determined that the way certain machinery was configured required workers to overextend when operating or maintaining their equipment. In response, team members put together a list of equipment and locations that were used regularly and difficult to reach. The list was given to another team focused on hazard mitigation, who worked with the engineering and maintenance department to make equipment changes. This other team provided written

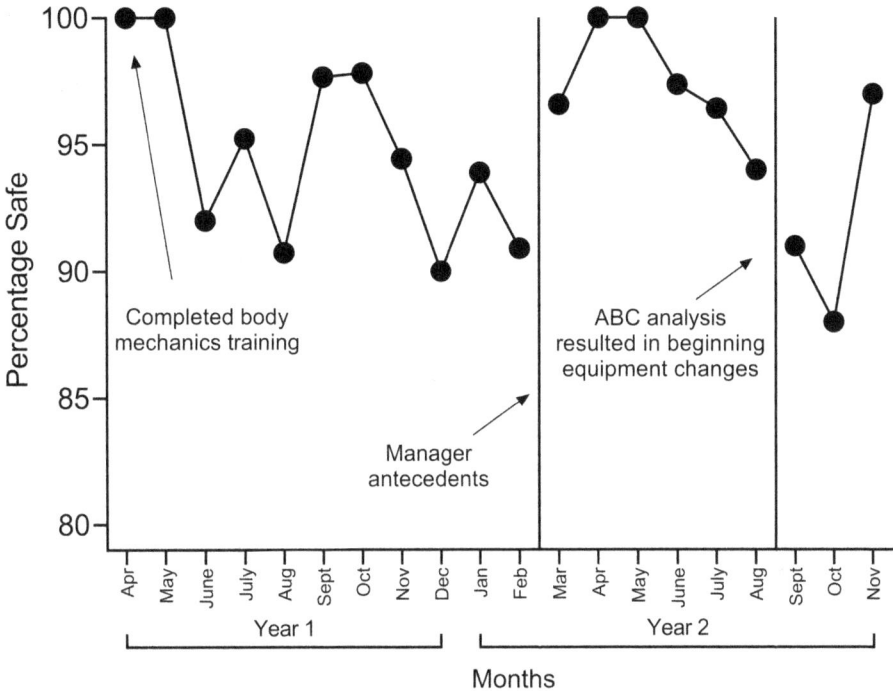

Figure 8.12   AWARE Program Overextending Safety Performance

Percentage safe behavior on the y-axis and consecutive months on the x-axis. The first phase change line indicates when manager antecedents were put in place. The second phase line indicates when an ABC analysis was conducted which resulted in equipment changes. Data are adapted with permission from the CCBS.

notifications updating the AWARE team of plans and actions aimed at getting the equipment fixed or adjusted. As of the time of writing, a number of changes had been made to equipment and overextension risk began showing positive change toward safe reaching (Figure 8.12).

Behavior Change Due to Peer Feedback

The peer observation and feedback process reflects a strong link between antecedents and direct consequences. When a new pinpoint is announced to the workforce, this safe behavior becomes the focus of both formal (using the behavioral checklist) and informal (seeing the behavior in passing) observations. Upon witnessing the safe or at-risk behavior, workers are shaped to discuss observations with their peer. This discussion provides workers with feedback, which serves as a powerful consequence. Safe behaviors that are promoted through these reinforcing consequences via feedback are more likely to happen more often.

In contrast, at-risk behaviors receive corrective feedback as a consequence. This feedback, if effective, reduces the at-risk behavior and should be paired with suggestions for a safe alternative behavior. When the worker goes on to exhibit the suggested safe behavior, further reinforcing feedback can serve to strengthen this new behavior. Of course, if the observation and feedback system is lacking in quantity or quality, the new behavior may not get reinforced and drift may occur. However, if reinforcement from feedback is sufficient, the new behavior may become part of the worker's repertoire, in which case no further shaping may be necessary.

Therefore, the process of pinpointing, observation, and feedback may be sufficient to change behavior. A recent example of this is provided by a Fortune 500 private grocery distribution company (Ludwig & Laske, 2021). In the warehouse setting, production is closely related to profitability: labor dollar costs must be offset significantly by the volume of food shipped to generate a profit. Employee production standards, engineered to optimize volume, thus create time-based goals for workers who can earn bonuses for exceeding the standards. Workers may also suffer pay reductions or negative employment outcomes for failing to meet the standards (Goomas & Ludwig, 2009; Ludwig & Goomas, 2009). Therefore, these standards are significant contingencies for a whole class of behaviors that help workers save time. Many of these time-saving behaviors also put them at risk. Behaviors such as stepping off a pallet jack (a motor vehicle with forks for a pallet in the back) while still moving and driving full speed through intersections when transitioning from one row of product to another were areas of risk that previously resulted in many injuries.

The employee behavioral safety team knew they could not immediately influence many of the at-risk behaviors in this hurried environment. They could, however, attempt to shape successive approximations while they worked with management to address some of the bigger issues relating to production quotas. As a first step, the team pinpointed "Brake to reduce speed" and "Honk at intersections" in their observation cards and feedback.

Baselines of these two behaviors were below 70% safe—a sign of good pinpointing finding risk. The pinpoints were then announced to the workforce along with the low baseline rate. Workers conducted observations and gave each other direct, individualized feedback. In addition, the team announced their goal to reach above 90% safe for three months and posted the trends for these behaviors in a common area frequented by employees. Such group-level feedback has been demonstrated to be an effective tactic to improve group performance (Ludwig et al., 2010; Ludwig & Geller, 1991, 1997; Stephens & Ludwig, 2005). By the third month, the percentage of time workers slowed their vehicle and honked at intersections topped 80%. Within four months, the employees had raised their speed reduction above 90% safe; and after three consecutive months they were able to celebrate their success and retire their goal (see Figure 8.13). Honking horns at intersections also showed a positive trend upward, approaching the goal (see Figure 8.14). These results suggested that the feedback process was effective in changing behavior.

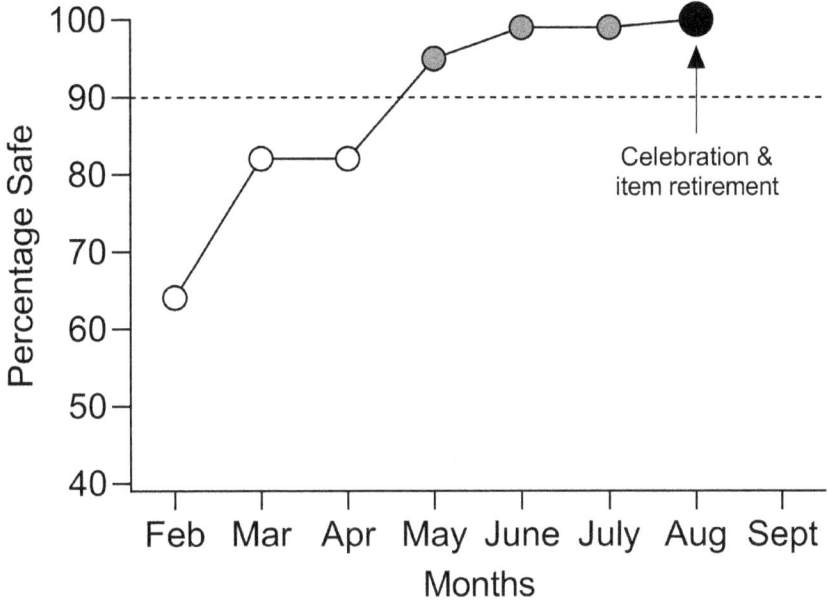

Figure 8.13   Reduce Speed at All Intersection Points

Percentage safe behavior on the y-axis and consecutive months on the x-axis. Gray circles indicate performance above the goal. The black circle represents when the checklist item met the three consecutive month criterion and was retired from the checklist.

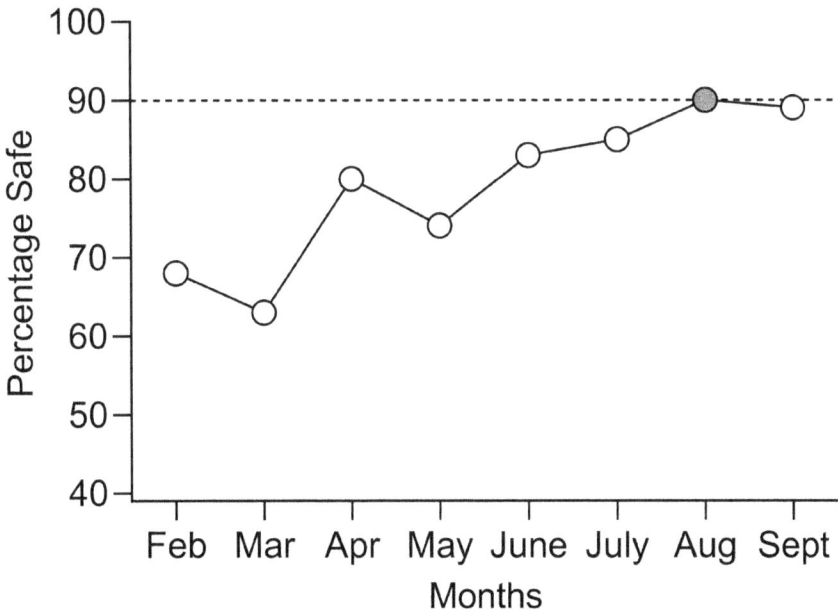

Figure 8.14   Sound Horn at All Intersections

Percentage safe behaviors on the y-axis and consecutive months on the x-axis. Gray circles indicate performance above the goal.

As at-risk behavior is publicly tracked, successful desirable behavior change can be a reinforcer for the workforce to target other behaviors. The behavioral safety program at SuperValu's Midwestern Region Distribution Center (MRDC) publicly posted pinpointed behavior trends in the employee lunch/breakroom. These data were displayed so workers could see positive trends toward their percentage safe goal. When the pinpointed behavior was above the goal (e.g., 90% safe) for three consecutive quarters, which often took years to achieve, the CAM team held a celebration with a free lunch to "retire" the pinpoint and adopt another.

This resulted in numerous examples of behavior change in the behavioral safety program (CCBS, 2013). One successful CAM pinpoint focused on workers who operate high-elevation forklifts that reach the high racks where surplus products are stored and replenish product cases on the lower racks where selectors pick their orders. After a baseline rate of 65% safe, operators increased their safe pulldown behavior to near goal levels (95% safe) within the year. However, it took them another year to successfully maintain safety performance over goal level and retire the pinpoint (Figure 8.15).

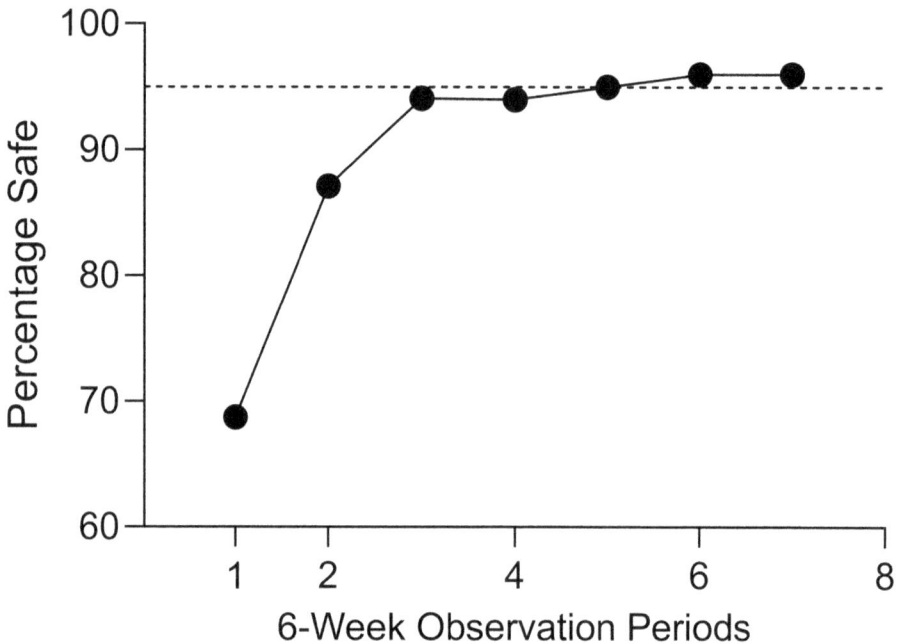

Figure 8.15   CAM Striking Pallets Pinpoint

Percentage safe behavior on the y-axis and six-week observation periods on the x-axis. Data were adapted with permission from the CCBS. From "Behavioral Safety: An Efficacious Application of Applied Behavior Analysis to Reduce Human Suffering," by T.D. Ludwig and M.M. Laske, 2022, *Journal of Organizational Behavior Management* (Taylor & Francis, 2022).

Similar success was achieved in relation to the upright placement of cases of liquids. When liquids are placed on pallets upside down, they can leak, causing slip hazards for workers and machinery in the warehouse, trucks, and grocery stores. Figure 8.16 shows that pinpointing and feedback were associated with immediate increases in this behavior, albeit not to goal levels. The workers achieved the 90% safe goal but were required to maintain this rate for three consecutive quarters to show mastery before retiring the pinpoint. The data dropped below the goal and they had to start over, ultimately achieving the goal and retiring the pinpoint.

Because of its success in safety, the CAM program also targeted behaviors related to quality in distribution operations. This new pinpoint helped selectors to fill orders more efficiently for more than one customer at a time. Selectors would open cases and take a couple of items (e.g., cosmetics) to put in a tote containing products fulfilling a customer order. When they put lettered clips ("ABC clips") on the totes, they were more likely to avoid putting the product in the wrong tote, thereby ensuring their customers (grocery stores) received only the products they had ordered. This behavior relating to the quality of operations

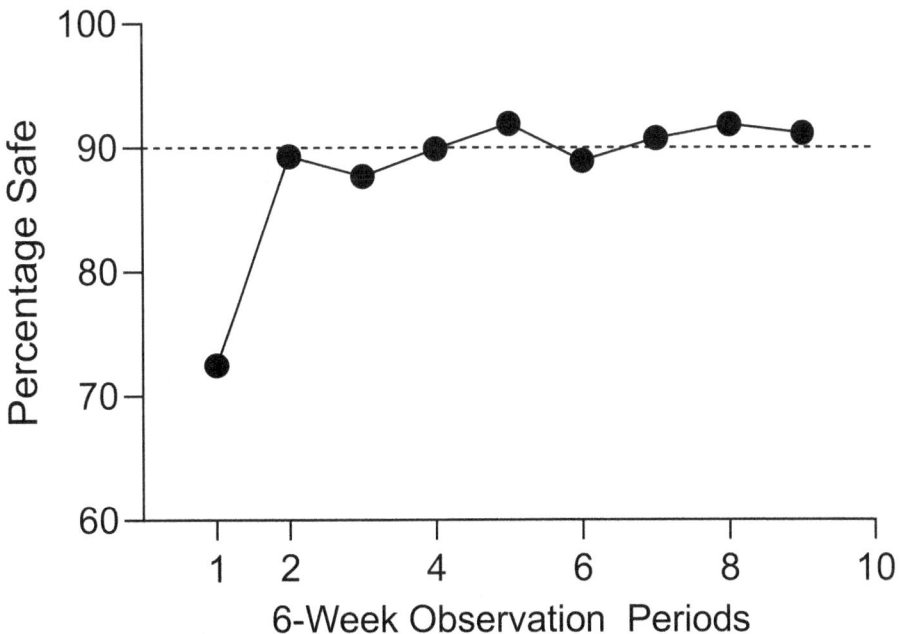

Figure 8.16   CAM Shipping Liquids Upright Pinpoint

Percentage safe behavior on the y-axis and six-week observation periods on the x-axis. Data are adapted with permission from the CCBS.

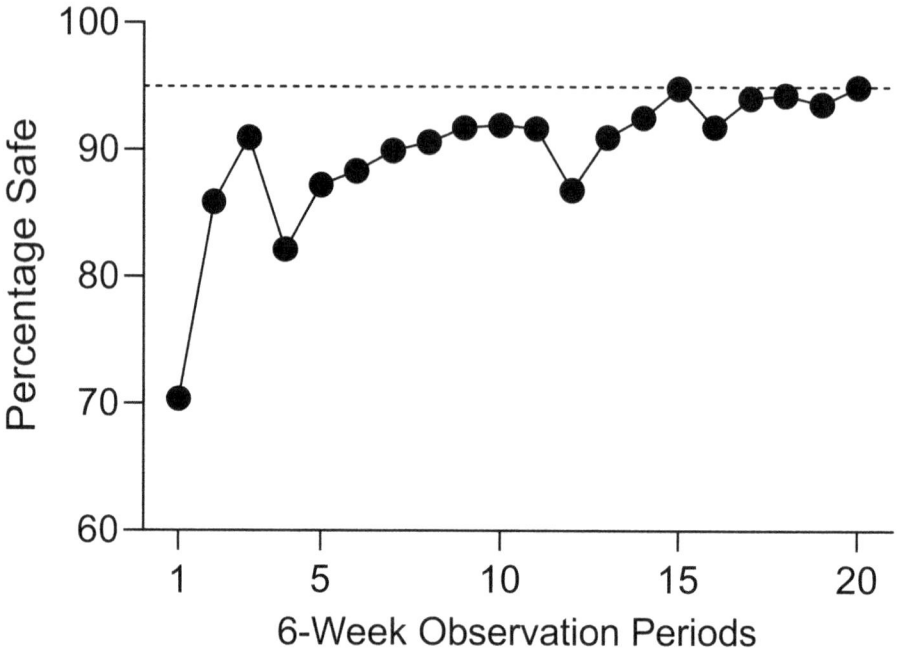

Figure 8.17   CAM ABC Clips Used as Required Pinpoint

Percentage safe behavior on the y-axis and six-week observation periods on the x-axis. Data are adapted with permission from the CCBS.

took nearly three years to achieve (Figure 8.17). After an immediate increase, the goal of 95% safe was not achieved until the CAM team implemented a "clip-it/ticket" program. The program rewarded associates with a ticket when observed clipping on the ABC clip properly. Draws were held in which an associate would receive a small incentive if their ticket was selected.

While three consecutive quarters above goal was the criterion to establish workforce mastery, the assumption was tested with follow-up observations. Turnover in warehouse labor could be as high as 50% in short timeframes. The team could not assume that new workers would adopt and sustain previously pinpointed safe behaviors. Therefore, the CAM team regularly conducted follow-up observations with special observation cards to revisit retired pinpoints.

Figure 8.18 shows a retired pinpoint related to safe lifting. Customer totes can get heavy and lifting them off the conveyer can cause risk. The CAM team adopted a pinpoint to slide the totes as a safe alternative to lifting. This pinpoint rose above the 90% safe goal and met the criterion after a year of observations. A year later, maintenance observation probes showed the behavior remained above goal. However, probes two years later showed the safe behavior may have fallen below goal levels. Therefore, the pinpoint was readopted, and the process successfully brought safe behavior back above goal levels. Nearly six years later, the CAM team found safe tote handling had dropped

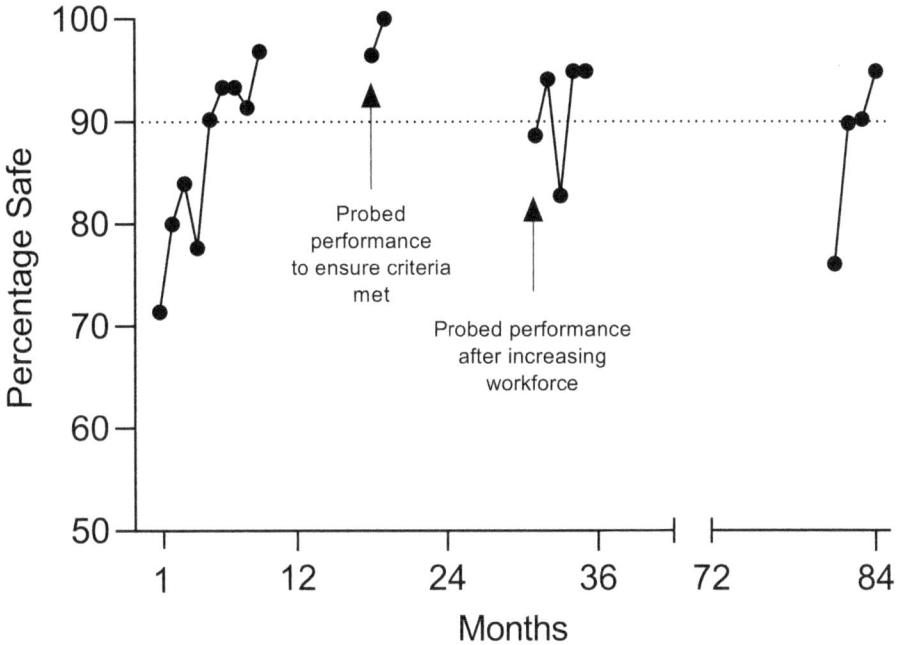

Figure 8.18   CAM Safe Lifting with Maintenance Probes

Percentage safe behavior on the y-axis and consecutive months on the x-axis. Data were adapted with permission from the CCBS. From "Behavioral Safety: An Efficacious Application of Applied Behavior Analysis to Reduce Human Suffering," by T.D. Ludwig and M.M. Laske, 2022, *Journal of Organizational Behavior Management* (Taylor & Francis, 2022).

back down to original baseline levels. The pinpoint was again adopted until the goal was achieved once again.

In both cases at the grocery distribution warehouse, no additional intervention was attempted on pinpointed behaviors beyond the basic processes of pinpointing and observation along with peer and group feedback. However, in some cases, behaviors pinpointed do not respond in a sustainable way to the direct individualized feedback inherent in most behavioral safety programs. In these cases, ABC analysis and business systems analysis are needed to identify changes to the contingencies surrounding the behavior.

On some occasions, simple antecedents may be a solution. Gribbins Insulation Company, which has a CCBS accredited program (CCBS, 2012d), had an observation process that was unsuccessful in identifying risks: monthly trends showed 100% safe across the board. In an attempt to impact this cultural practice of hiding risk, the team revised their pinpoint list to include "Working in adverse weather"—something that is very risky for insulation workers in a refinery working at height on scaffolding. It was also something that workers did not want to do. Therefore, increases in at-risk observation on this pinpoint occurred over the winter and spring months. After documenting that roughly

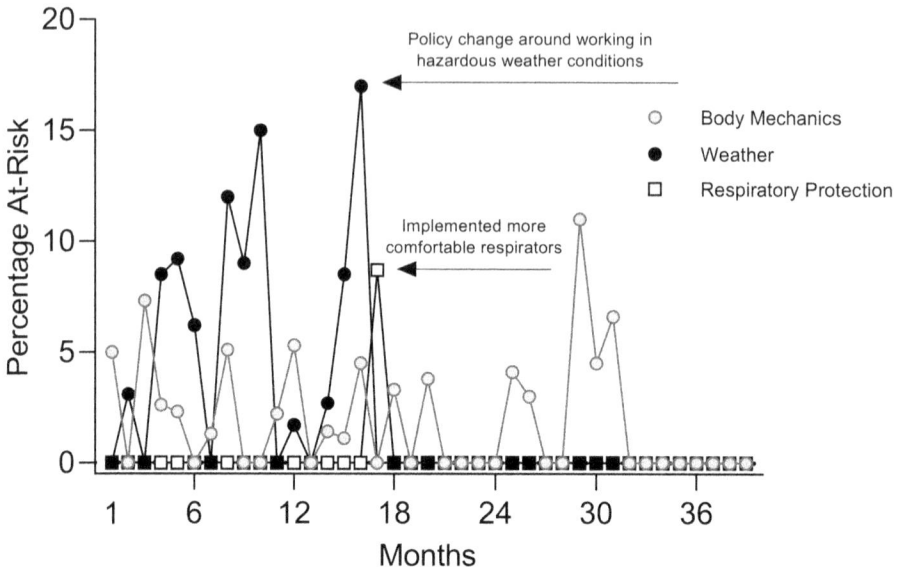

Figure 8.19 Gribbins Insulation Behavior Change Interventions

Percentage safe behavior on the y-axis and consecutive months on the x-axis. Black circles indicate at-risk behavior related to weather. White squares indicate at-risk behavior related to respirators. Gray circles indicate at-risk behavior related to body mechanics. Data are adapted with permission from the CCBS.

20% of observations found that working in weather was related to risks, the team showed their data to the safety manager and owner of the company. This meeting led to a policy statement allowing workers to stop work when certain hazardous types of weather are imminent. This simple policy, based on data, successfully dropped risks associated with working in adverse weather to zero (see Figure 8.19).

Once the workers experienced how their observations could lead to improvements in safe working, other pinpoints that had previously been reported as 100% safe started being reported as at-risk. The month after the weather policy was implemented, a rise in at-risk behaviors related to respiratory protection was observed. Undoubtedly, workers were now more likely to record at-risk behavior because they had experienced positive changes from identifying risk in other observations. Most likely, the at-risk behaviors noted in observations such as "Pulling respirators to the side or on top of the head" had been happening previously, but workers had not wished to disclose them. ABC analysis revealed that the respirators were uncomfortable and did not secure well, leading workers to pull them out of the way so as not to distract from their tasks. Building on their success with the weather policy, the team took this new data to their managers, who then brought in respirator suppliers and let the team choose the best-fitting, most comfortable respirators for use by the company's workers. With the new respirators, the at-risk behaviors related to respirators dropped to zero.

PERFORMANCE

A behavioral safety program operating with integrity, producing quality observations and feedback causing behavior change, should result in enhanced safety performance and injury reduction. If this doesn't happen, we need to consider if behavioral pinpointing is failing to target emerging risks or if the behavioral program is not garnering sufficient support in the workplace culture. Indeed, as Tom Gilbert noted in his seminal work *Human Competence: Engineering Worthy Performance* (1978), behavior is costly (and behavioral programs can be costly), but accomplishments are worth investing in. He offers the formula $W = A/B$ to suggest that Worthy Performance (W) is a function of valuable Accomplishments (A) divided by the costs associated with Behavior Change (B). Therefore, the costs of behavior change and associated behavioral programs must result in measurable and impactful accomplishments. This section reviews the empirical evidence for behavioral safety program performance by documenting areas of notable accomplishments.

Leading Indicators

One important yet often overlooked accomplishment of behavioral programs is the impact on other safety management systems. Think about it: all safety management systems engaged in by a company are indeed behavioral programs, because they seek to impact worker safety behavior and/or get workers to report issues such as the physical hazards that they come in contact with. Safety training programs seek to build workers' safety repertoire; standard operating procedures, instructions, and rules offer antecedents; discipline and reward programs attempt to provide consequences; observation, inspections, and audits seek to provide feedback; lockout/tagout and permitting provide processes designed to manage behavior; and supervision and safety meetings attempt to deliver frequent discriminative stimuli.

In most behavioral safety programs, the direct observation of behavior results in the identification of risk when this data is submitted and collected. One might consider the behaviors involved related to "Reporting safety issues." Reporting is a behavior! Most safety programs have other mechanisms for workers to report safety issues. These include participating in inspections; making safety suggestions on a form or to a supervisor; and reporting near misses when hazard energy is released but does not strike the body (a.k.a. "close calls") and minor injuries that don't require the worker to seek medical attention or get a prescription.

As workers' behavioral observations are reinforced through program success, one might expect to see some response generalization[SM-8.1] (DeRiso & Ludwig, 2012; Ludwig, 2002; Ludwig & Geller, 1991, 1997) to other "reporting" behaviors. In other words, as behavioral observations increase, we would expect to see worker participation in safety management systems increase, along with reporting of near misses and minor injuries.

Science Moment 8.1
Response Generalization

"Generalization" means that your behavior change shows up in some other way than was targeted. "Stimulus generalization" occurs when responses reinforced in one setting generalize to other settings. For example, if you shape maintenance workers at a dam in the behaviors necessary to safely replace turbine blades in a simulated training context and then observe that they were also able to do this successfully in the confined turbines in the plant, this constitutes stimulus generalization. "Response generalization" occurs when you shape one behavior (e.g., checking pressure gauges before decoupling a hose) and then observe similar but different behaviors (e.g., standing to the side of the hose opening) starting to happen on their own.

A classic example of response generalization in safety is set out in an article by Ludwig and Geller (1997). While reinforcing "complete intersection stopping" in pizza delivery drivers, they also observed increases in nontargeted behaviors (e.g., "turn signal use" and "safety belt use"). However, response generalization was only observed in the groups of pizza delivery drivers who had input in the pinpointing and goal-setting process—yet another reason to involve the front line in pinpointing.

After Eastman Chemical's AFD behavioral safety program (CCBS, 2015a) had been operational for four years, the AFD instituted voluntary near miss reporting. Initially, near miss reporting had low levels of participation, averaging about 50 reports of near misses a year from a workforce of over 400. The number of near miss reports doubled and then nearly tripled over the subsequent seven years. Substantial increases continued to around 400 near miss reports as AFD engaged in enhancements to its behavioral program. These improvements included migrating from mandatory to voluntary observations; a new employee mentoring program where new workers were trained in and practiced behavioral observations (among other competencies); and a "shared learning" program where safety professionals shared the causes of significant injuries/near misses with the workforce, asking the behavioral safety team to target related pinpoints (see Figure 8.20).

As near miss reporting increased, the number of injuries within the AFD was so low (ORIR of less than 0.2) that safety professionals realized they were no longer doing incident investigations. They decided to analyze the "high

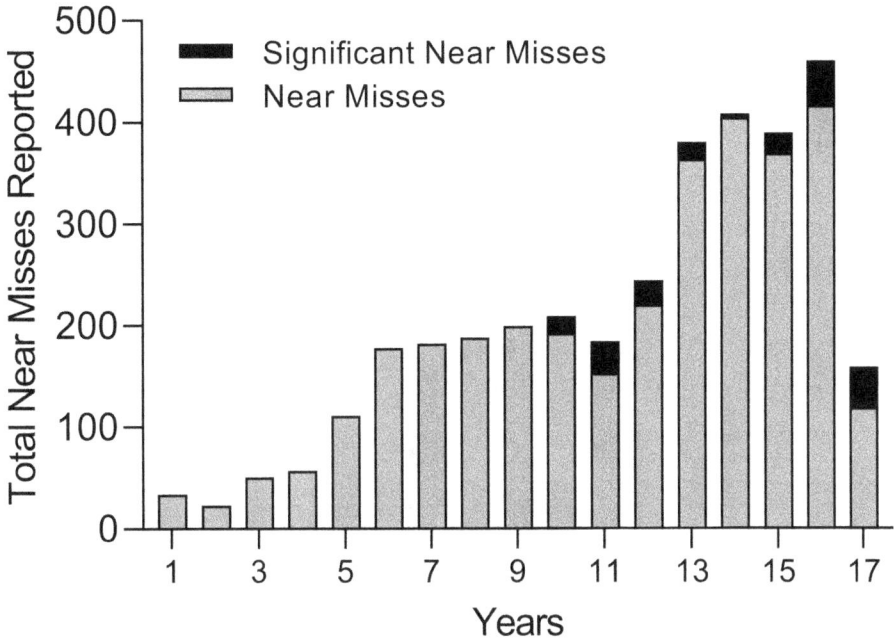

Figure 8.20 AFD Near Misses Reported

Near misses reported on the y-axis and consecutive years on the x-axis. Near misses are indicated in the gray bar. Significant near misses are in the black bar. Data are adapted with permission from the CCBS.

potential" near misses that could have resulted in a serious injury or fatality as if these had been actual serious incidents. This allowed them to "stay in practice" for their investigation skills and, more importantly, proactively address emerging safety issues by acting on their analysis before an actual injury was sustained.

Eastman Chemical also tracked workforce participation in a new task safety audit program, where workers teamed with managers and safety professionals to conduct a thorough task assessment focusing on behaviors, process issues, hazards, and safety equipment performance. These audits assist in removing hazards from the work area and provide information to improve training, processes, and safety equipment. Only 138 task safety audits were conducted over the course of the second year of the corporate program by the 400 workers in the AFD (which was a higher percentage than in other company divisions). The audit data became an excellent source of risk analysis for the behavioral safety team, who began to adopt new pinpoints based on the findings. Because the workers were familiar with behavioral observations and because they recognized the synergy between these programs, every worker in the division agreed to do at least one audit a year, resulting in a nearly 200% increase (see Figure 8.21).

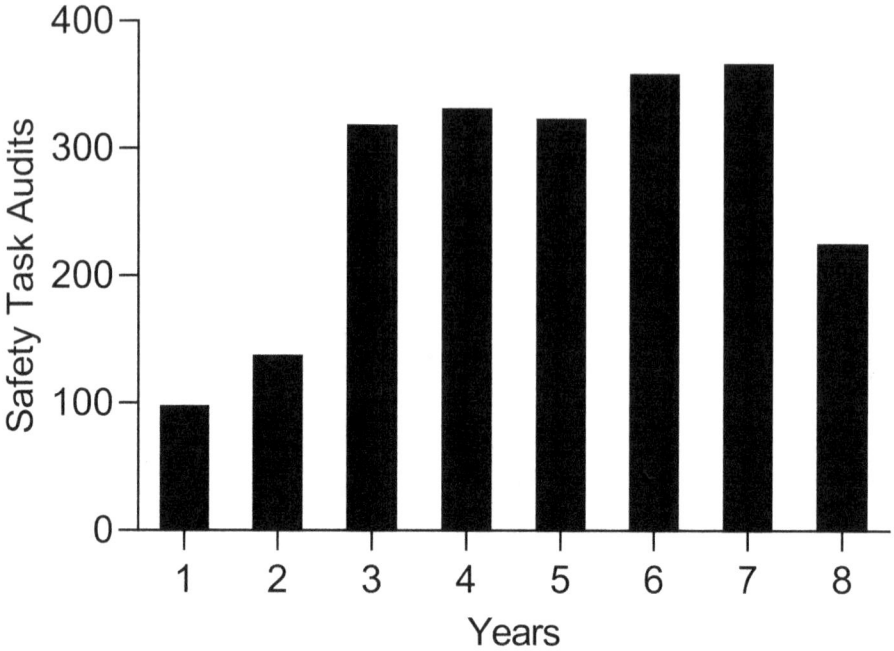

Figure 8.21   AFD Task Safety Audits per Year

Safety task audits are on the y-axis and consecutive years on the x-axis. Data are adapted with permission from the CCBS.

Reduction in Injury

Injury reduction is the primary purpose of behavioral safety. While there have been many detractors who suggest that behavioral safety does not work, blame workers for their behavior, or do not consider environmental influences[1] (Frederick & Lessin, 2000; Howe, 2001; Mathis, 2009; Metzgar, 2011; T.A. Smith, 1999), the CCBS Commission on Behavioral Safety has been documenting the effectiveness of behavioral safety programs in reducing injuries since 2005 (CCBS, 2022b). Programs must demonstrate a reduction in injuries associated with the onset and/or enhancement of their behavioral safety programs. Programs must show not only substantially lower injuries below baseline levels, but also that injury rates have been lower than the industry standard for a period of three years. These stringent criteria were put in place to differentiate the results from typical injury rates reflecting industry's best safety practices.

To date, 23 behavioral safety programs have been accredited by the CCBS, with many reaccredited multiple times. Table 8.1 displays average injury rates for each CCBS accredited program five years before and after implementation. Reductions in injury rates are demonstrated across all programs where data were available. These findings indicate robust relationships

Table 8.1 CCBS Accredited Program Injury Rates Five Years Before and After Behavioral Safety Implementation

Company	Five-year average ORIR/accident frequency rate (AFR) rate before behavioral safety program	Five-year average ORIR/AFR after behavioral safety program
Acetate Fibers Division, Eastman Chemical Company	16	4.78
Ahlstrom-Munksjo Italia	(4.36)	(0)
Bay Industrial Safety Services	Not available	Not available
Brand Energy Services	Not available	0.08
Costain	(0.21)	(0.14)
Freitag-Weinhardt	Not available	Not available
Gribbins Insulation Company	1.02	0
Halliburton Gulf of Mexico	1.79	1.12
Illinois Refining Division, Marathon Petroleum Company	3.39	2.05
Lytle Electric Company	2.78	0
Michigan Refining Division, Marathon Petroleum Company	Not available	Not available
Midwest Regional Distribution Center, SuperValu	22.22	4.47
Mistras Group	Not available	0
Morris Construction	Not available	0.29
Ohio Refining Division, Marathon Petroleum Company	Not available	Not available
SDR Coating	2.91	0
SENCO Construction	Not available	0.47
Southeast Regional Facility, SuperValu	13.77	2.78
St. Paul Park Refinery, Marathon Petroleum Company	Not available	Not available
Stewart Security Patrol Incorporated	Not available	0.20
Texas Refining Division, Marathon Petroleum Company	Not available	Not available
Western Energy Company	6.61	1.92
White Construction	Not available	Not available

Data are representative of each accredited site's average injury rate five years prior to and after implementation of a behavioral safety process. The first year of behavioral safety implementation is not included in the average. Unavailable data are due to program implementation before data were reported in each CCBS application.

between implementation of a behavioral safety process and injury reduction, regardless of the year of implementation or industry.

The programs in Table 8.1 have received accreditation based on the principles of behavior analysis, which not all "behavior-based safety" (BBS) programs follow. "BBS" is simply a marketing moniker used by consultants trying to sell services. Not all deliverers of BBS follow or even know the science of behavior analysis. Therefore, these results may not reflect all BBS implementations, in that some lacked success due to the absence of the scientific principles outlined in this book.

Perhaps the most robust demonstration of the impact of a behavioral safety program was in the injury data from SuperValu (CCBS, 2013). The CAM program was first implemented in SuperValu's MRDC. The MRDC plant experienced a substantial downward trend, eventually having a year without an injury (unheard in the food distribution industry), until significant product volume, staffing and overtime increases (in Year 10) increased injury rates to industry standards for a year.

Because of this success, other distribution centers in SuperValu's system were encouraged to adopt the CAM behavioral safety program. A second distribution center—the Southeast Regional Facility (SERF) (CCBS, 2012a)—adopted CAM about five years after MRDC. After implementing the CCBS accredited program, SERF also showed an immediate downward trend in injuries, replicating MRDC's success in a natural multiple baseline[SM-8.2] (Figure 8.22; see Erath et al., 2021 for a review of experimental methodologies in organizational behavior management). This replication gives us a high degree of confidence in the relationship between behavioral safety and injury mitigation over and above other safety initiatives in which the company may have been engaged.

Western Energy is a 25,000-acre coal mining operation in Montana with about 300 employees. Western Energy's injury reporting to the Mining Safety and Health Administration averaged six recordable injuries per 100 employees (roughly 18 injuries a year). An employee design team facilitated the BESAFE program (CCBS, 2015b). After implementation, injury rates fell 30% over two years and dropped another 30% to an annual rate of 2.2 per 100 full-time equivalent workers. Five years after BESAFE started, the mine produced a year of no recordable injuries. It sustained an injury rate below the mining industry standard, returning to zero the year it was accredited by the CCBS (see Figure 8.24).

Ahlstrom-Munksjo Italia is a group of two plants in Italy producing specialty paper for liners and filtration for industrial uses. The plants employ about 500 employees (one plant is three times larger than the other). After a decade of decreasing injury rates—measured in Europe as the number of incidents (three days off work) per million labor hours (i.e., accident frequency

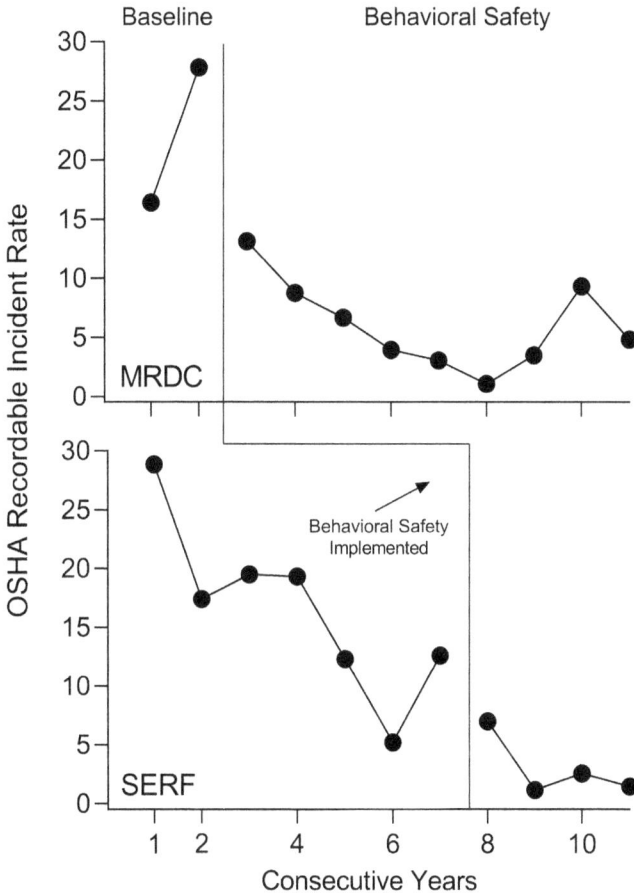

Figure 8.22   MRDC and SERF Multiple Baseline Across Distribution Centers

ORIR on the y-axis and consecutive years on the x-axis. Phase change lines indicate when behavioral safety was implemented in the respective distribution centers. Data were adapted with permission from the CCMS. From "Behavioral Safety: An Efficacious Application of Applied Behavior Analysis to Reduce Human Suffering," by T.D. Ludwig and M.M. Laske, 2022, *Journal of Organizational Behavior Management* (Taylor & Francis, 2022).

rate (AFR)—the plants' injury rates fluctuated between annual rates of 1.5 and 6.61. The implementation of an incentive program did little to impact their AFR. However, after the implementation of their behavioral safety program (CCBS 2018a), the two plants performed without any recordable injuries for the three years leading up to accreditation (the industry standard was nearly 20.0), with zero lost-time injuries (see Figure 8.25). Notably, the introduction of behavioral safety at Ahlstrom-Munksjo Italia was also associated with a 240% increase in near miss reporting.

Science Moment 8.2
Multiple Baseline

A "multiple baseline" design is an experiment to see if our intervention is indeed what caused the change in behavior or injuries, or if there may have been other causes (i.e., history effects). In an industrial setting, there is a lot going on in terms of efforts to reduce injuries and increase safe behaviors. How do you know if any changes are the result of your behavioral program or some other corporate initiative? The key feature of a multiple baseline design is the staggered introduction of the intervention across time and groups (e.g., a department). Consider the example provided in Figure 8.23. Here we have two departments in Part A, where the company intervened with a bonus system to increase the number of parts produced. Department 1 got the intervention first, followed by Department 2 about 14 days later. After the intervention was introduced in Department 1, performance increased; while performance in Department 2 remained the same during that time, giving us a higher degree of confidence in the positive impact of the intervention. Contrast this data with Part B, where Department 1 again increased production. However, productivity also increased in Department 2 even though the intervention had not yet been introduced. Because the changes happened in Department 2 while it was still in baseline, before the intervention was introduced, something else must have impacted productivity. Therefore, it would be incorrect to assume the intervention, in and of itself, increased production.

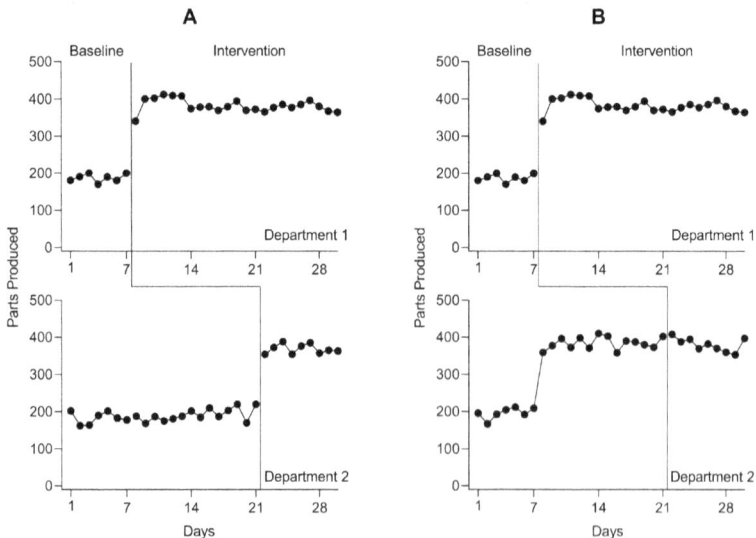

Figure 8.23   Multiple Baseline Design

Number of parts produced on the y-axis and consecutive days on the x-axis. Phase lines indicate when the intervention was put into place.

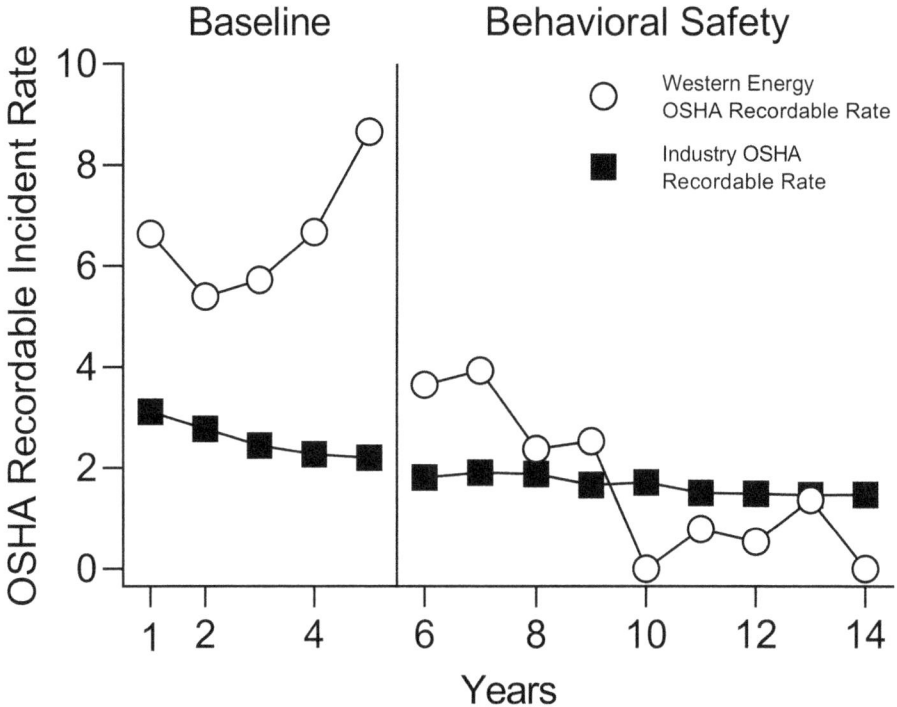

Figure 8.24 BESAFE ORIR Comparted to Mining Industry Standard

ORIR on the y-axis and consecutive years on the x-axis. White circles indicate Western Energy's ORIR. Black squares indicate the industry standard ORIR. The phase line indicates when behavioral safety was implemented at Western Energy. Data were adapted with permission from the CCBS. From "Behavioral Safety: An Efficacious Application of Applied Behavior Analysis to Reduce Human Suffering," by T.D. Ludwig and M.M. Laske, 2022, *Journal of Organizational Behavior Management* (Taylor & Francis, 2022).

Costain —a large construction engineering and general contracting company in the U.K.—engaged in a variety of behavioral safety programming to reduce its AFR by 600% (see Figure 8.26; CCBS, 2014b, 2018b) to a level representing just one injury (three days off work) for every million labor hours, while increasing its worker population by 67% (CCBS, 2014b). Costain's unique behavioral safety program, where all management are trained in behavioral science by an internal team of behavior management experts, is paired with an observation program, where workers and supervision annually report over 30,0000 safe behaviors, hazards, and near misses, all on one form.

Marathon Petroleum Company's Refining Division boasts five refineries with behavioral safety programs accredited by the CCBS (Illinois Refining Division [CCBS, 2019]; Michigan Refining Division [CCBS, 2020a]; Texas Refining Division [CCBS, 2017a]; Ohio Refining Division [CCBS, 2016];

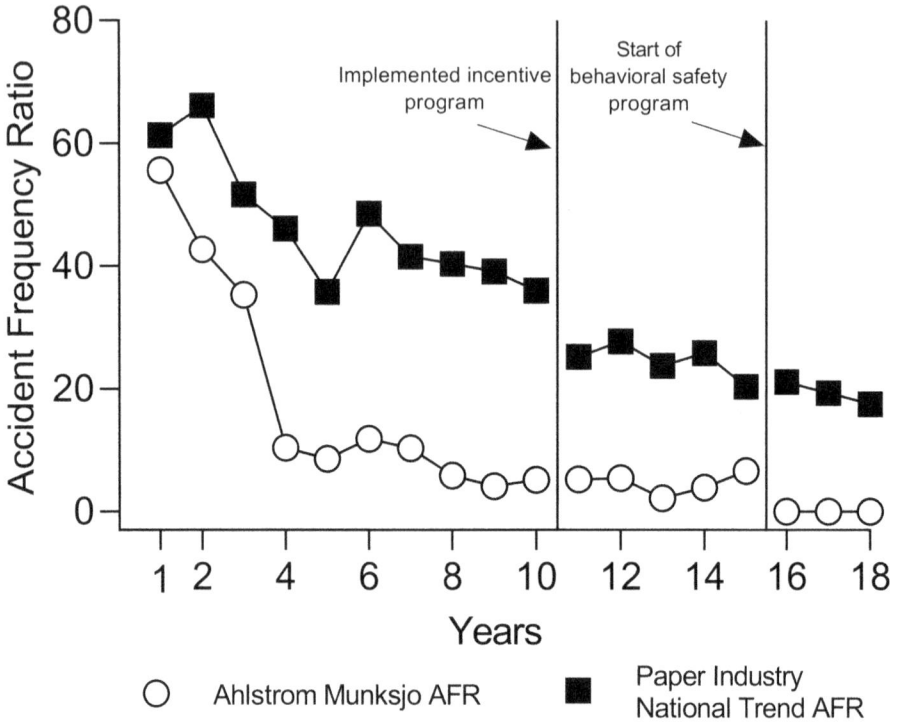

Figure 8.25   Ahlstrom-Munksio Italia AFR and Industry Standard AFR

AFR on the y-axis and consecutive years on the x-axis. White circles indicate Ahlstrom-Munksjo's AFR. Black squares indicate the industry standard AFR. Data are adapted with permission from the CCBS.

St. Paul Park Refinery [CCBS, 2020b]). Each showed substantial and sustained decreases in injuries with the onset of its behavioral safety programs. Figure 8.27 demonstrates the relationship between the performance of these behavioral safety programs, measured by participation in behavioral observations, and the real impact on injury rates.

SPREAD OF EFFECT

As leaders, supervisors, and work teams shape their behavioral skills through the practice of behavioral safety and its reinforcing outcomes (e.g., reductions in injuries), we may expect that other organizational results may be impacted through these behavioral skills. Communication processes are improved by interactions between employees and management through multiple behavioral safety components. Indeed, once the employee behavior of reporting safety concerns is reinforced, one may expect that this response will generalize to other types of reporting of issues impacting service and product quality,

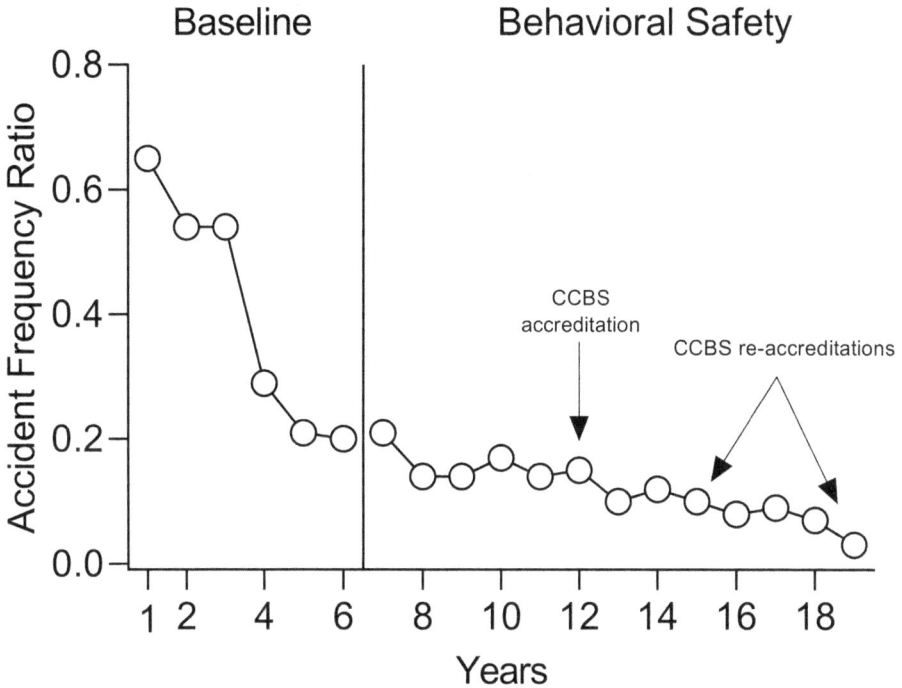

Figure 8.26 Costain AFR and Accreditations

AFR on the y-axis and consecutive years on the x-axis. Phase line indicates when behavioral safety was implemented at Costain. Arrow lines indicate when Costain received (re)accreditation from the CCBS. Data are adapted with permission from the CCBS.

process quality, workflow problems, staffing implications, cross-functional inefficiencies, material shrink and waste, security threats, and the myriad of other factors that—in large and small ways—impact organizational performance (DeRiso & Ludwig, 2012; Ludwig, 2002; Ludwig & Geller, 1991, 1997; and see Science Moment 8.1). Improvements in tools and equipment necessitated by ABC analysis to reduce risk can be associated with better production overall. Processes and systems can be brought under control (Deming, 1982, 1986) through the successful application of behavioral systems analysis. As work processes are analyzed and improved to reduce the occurrence of at-risk behavior, other behaviors that impact quality, production, and work culture may also be affected; in fact, these may be the same behaviors and/or response classes as pinpointed in the behavioral safety program.

Certainly, more research is needed to affirm this claim. However, there is evidence of this spread of effect across the CCBS accredited programs associated with reductions in injury. At the Marathon Petroleum refineries, discussed above, the reduction in injury associated with behavioral safety programming

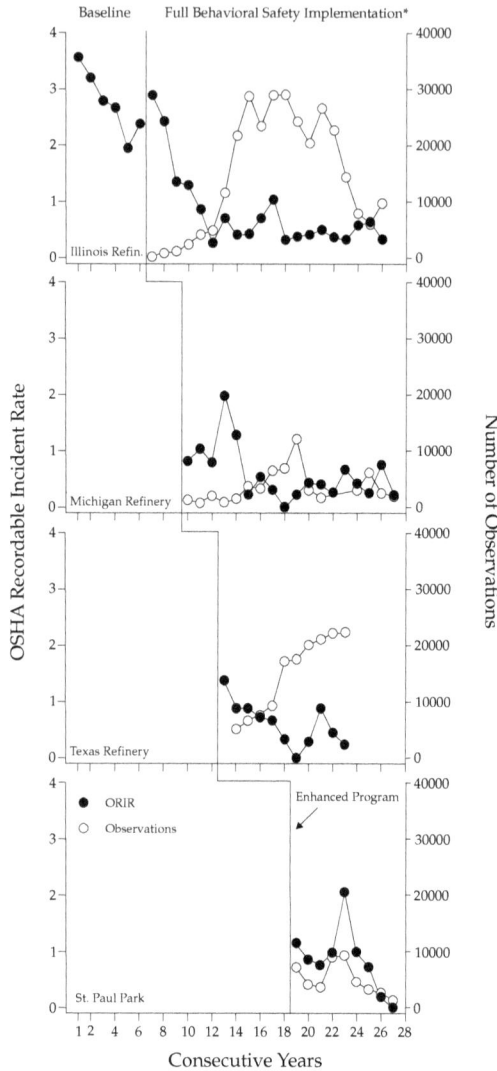

Figure 8.27 Marathon Refinery Sites: ORIR and Observations

ORIR on the primary y-axis; number of observations on the secondary y-axis; and consecutive years on the x-axis. Black circles indicate ORIR. White circles indicate number of observations. Phase lines indicate when behavioral safety was fully implemented.
The criteria for full implementation included:

- dates and/or data reported to the CCBS Commission on Behavioral Safety;
- full implementation after pilots; and
- programmatic enhancements (e.g., the establishment of a full steering committee).

Some refineries did not provide baseline data prior to implementing the behavioral safety program. However, they still reported decreases in ORIR after implementation of their behavioral safety program. Data were adapted with permission from the CCBS. From "Behavioral Safety: An Efficacious Application of Applied Behavior Analysis to Reduce Human Suffering," by T.D. Ludwig and M.M. Laske, 2022, *Journal of Organizational Behavior Management* (Taylor & Francis, 2022).

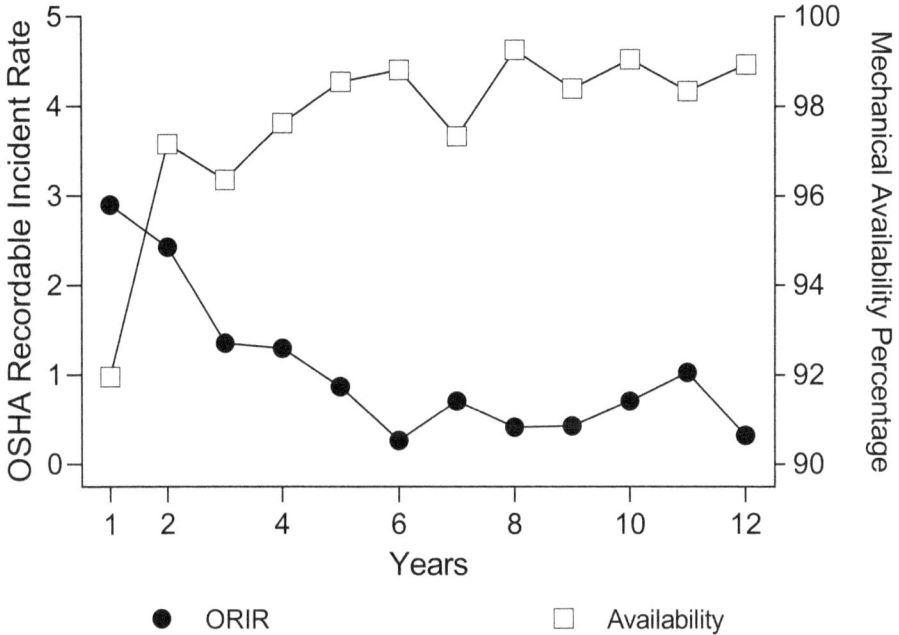

Figure 8.28   Marathon Illinois Mechanical Availability and ORIR

Mechanical availability percentage on the primary y-axis; ORIR on the secondary y-axis; and consecutive years on the x-axis. Black circles indicate ORIR. White squares indicate percentage of mechanical availability. Data are adapted with permission from the CCBS.

seems to relate to increases in mechanical availability—a measure of refineries' processing unit capacity to operate without impacting other operations or the quantity/quality of the end commercial product (see Figure 8.28; CCBS, 2012b). This finding may suggest that maintenance crews who are responsible for mechanical availability improve their reliability performance along with their safety performance. A bit of conjecture might suggest that the honest, no-blame peer-to-peer feedback shaped when participating in behavioral safety may also be used by workers to increase the quality of work behaviors that impact reliability.

A similar finding was discovered in Eastman Chemicals' AFD when it experienced decreases in customer product complaints as it decreased injury rates through engaging in behavioral safety programming (see Figure 8.29). Customer complaints are much like injuries in that they are a lagging indicator of performance, with errors in processing sometimes related to behaviors that put the product at risk. It is possible that the same behaviors that put employees at risk may put the product at risk. Perhaps as employees and managers addressed the at-risk behaviors that led to personal injury, the resulting improvements in communication, processes, and systems could have also impacted product quality.

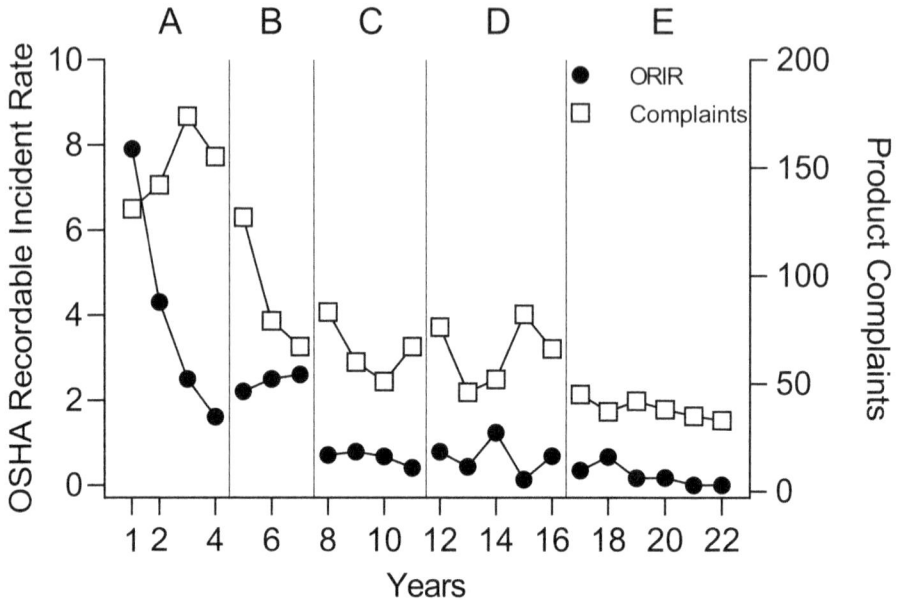

Figure 8.29 AFD ORIR and Customer Product Complaints

ORIR on the primary y-axis; product complaints on the secondary y-axis; and consecutive years on the x-axis. Black circles indicate ORIR. White squares indicate product complaints. At phase A, the behavioral safety program was implemented. Phase B indicates full implementation of the behavioral safety program. Phase C indicates division restructuring and program improvements. Phase D indicates first program accreditation from the CCBS. Phase E indicates program reaccreditation and improvements based on CCBS recommendations. Data are adapted with permission from the CCBS.

There is a common assumption that productivity (i.e., how much product is produced) is negatively correlated with safety performance. As production increases due to customer demand or increased profit goals, typically the performance of behavior must increase to meet the requirement. As workers hurry to get tasks done, they may omit behaviors that keep them safe and/or engage in at-risk behaviors that save time. Workers may also experience more fatigue, which erodes behavioral fluency, putting them at risk. Therefore, it is also often assumed that increases in safety behavior will harm productivity. However, there are examples across the CCBS accredited programs of the opposite effect. In some cases, decreases in injury rates were associated with productivity increases. For example, as Costain decreased its injury rates, it experienced a 33% increase in the number of construction projects awarded and a further increase in workers employed (CCBS, 2018b).

Probably the best example of how a hurried work process can lead to injuries is in the distribution industry, where selectors and replenishers are incentivized to meet or exceed production standards. While these standards are engineered to be within 80% of human capability, the incentives nonetheless reinforce behaviors that enhance worker production. Workers who fail to meet the standards consistently may lose their jobs. This negatively reinforces less-skilled workers to engage in risky behaviors to keep up with the standards. Workers with greater fluency can earn monetary bonuses by exceeding the standards, leading to at-risk behaviors among others who are not as fluent but are still trying to earn the bonuses. In contrast, to avoid injury on these jobs, workers must engage in behaviors that may slow them down (e.g., coming to a complete stop at intersections and when they step off their pallet truck to retrieve a case). Leaders of distribution companies may consider avoiding behavioral safety programs that decrease injuries because they believe these could also decrease productivity.

However, at SuperValu MRDC (CCBS, 2013), an inverse relationship was found between injury reduction associated with the behavioral safety program and productivity. MRDC increased both safety performance and productivity through its behavioral safety program (see Figure 8.30). After introducing the behavioral safety program, MRDC experienced an 80% decrease in injuries over a six-year period. During that same time, MRDC's production (i.e., cases picked per hour) increased 25%. Corporate executives visited the site to attend

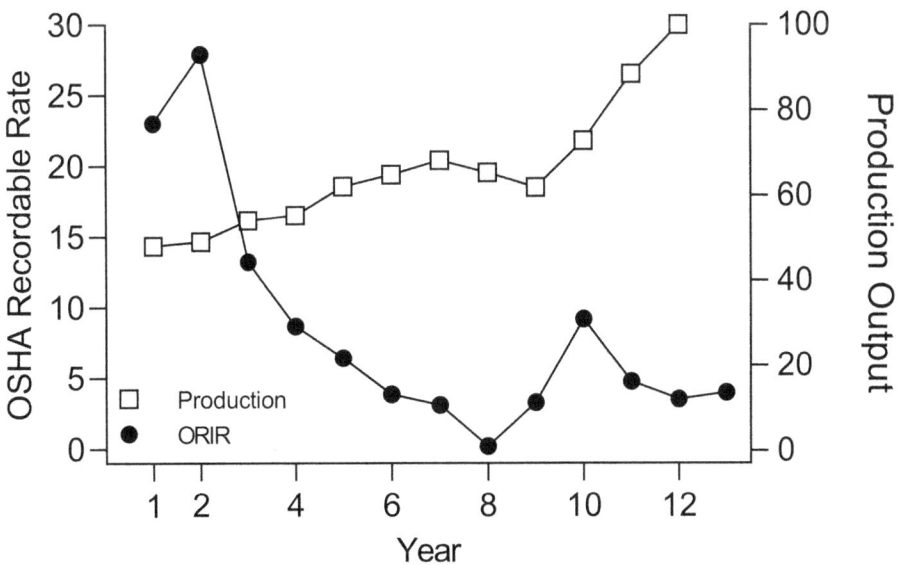

Figure 8.30   MRDC ORIR and Production Output

ORIR on the primary y-axis; production output on the secondary y-axis; and consecutive years on the x-axis. Black circles indicate ORIR. White squares indicate production output, measured as the number of outbound cases per hour dived by the number of production hours. Data are adapted with permission from the CCBS.

the accreditation celebration, where they decided to deploy the CAM process across the company after seeing this data.

The negative relationship between production and safety reared its ugly head, however, when the corporation bought another grocery retailer, closed a nearby distribution center and then integrated that larger distribution center's volume into MRDC's production. This resulted in a surge in injuries, as the significant increase in volume doubled the workforce (having to use unskilled labor) and overtime. As noted earlier, the CAM process had to respond by increasing observations and, within a year, injury rates had returned below the industry average while production continued to rise.

Caveat

These repeated demonstrations of injury rates decreasing after the onset of behavioral safety programming are presented as evidence of the effectiveness of the science. However, companies typically do not use just one tactic to improve performance in safety or other areas. Therefore, we must be considerate of history effects (Campbell & Stanley, 1963) in the real world, where other safety programming and operational decisions may also impact injury reduction (see the referenced CCBS citations on behavior.org which include other safety programming notes). While we argue that all safety management systems are behavioral interventions, we cannot segregate the impact of one while others are being implemented and/or improved during the same period. Additionally, other business changes in production during these periods can have an impact on injuries. Leadership improvements, new equipment, and growing fluency within a tenured workforce can all contribute to injury reduction. Therefore, we can only note that decreases in injury rates are *associated* with behavioral safety programming, not necessarily *caused* by it.

Similarly, correlation does not assume causality. The observed relationships presented here between injury and participation rates, leading indicators, and other organizational performance metrics (reliability, quality, production) are just correlations and must be interpreted with added caution. We do not know if safety performance, due to many potential factors, impacted other areas of performance (e.g., mechanical reliability, customer satisfaction, production); or if the company's actions to improve other areas of performance also had the effect of reducing injuries. For example, to increase mechanical availability in a refinery, the maintenance staff must engage in more and higher-quality preventive maintenance and reactive maintenance tasks. As a result, the equipment becomes less hazardous for workers because there is less chance of unplanned energy release and less downtime when workers must engage in riskier variant behaviors.

We must also consider the third variable, where an event impacts both safety and other areas of organizational performance. Certainly, leadership shake-ups can bring in a more competent or risky executive who makes sweeping changes in budget, production goals, and/or product. Something behavioral systems analysis teaches us is that when one system is out of control, it is difficult to bring other systems under control. If upstream planning, financing, HR engineering, procurement, and other processes are variable in their performance, it is an uphill battle for workers to avoid risk. Where a company intervenes to get its systems under control via leadership changes, IT solutions, or process improvement programs like LEAN (Holweg, 2007; Shah & Ward, 2007), Six Sigma (Swink & Jacobs, 2012; Zu et al., 2008) or International Organization for Standardization certification (Boiral et al., 2018; Heras-Saizarbitoria & Boiral, 2013), we would also expect improvements in safety, which depends on stable and effective behavioral systems.

NOTE

1. We find it interesting that critics of behavior analysis, more broadly, charge the field with environmentalism (Mahoney, 1989; see Todd & Morris, 1992 for a review)—incorrectly of course (Catania, 1991; Skinner, 1974); while behavioral safety is misrepresented as not considering environmental influences (DeJoy, 2005; Howe, 2001), albeit for different reasons.

CONCLUSION

SETTING THE RECORD STRAIGHT

Reggie was excited to share the success of their behavioral safety program with other plants in the company. However, it was not all sunshine and rainbows elsewhere. In fact, when Reggie spoke to participants in other behavior-based safety (BBS) programs, he easily identified that what they were doing was NOT based on the science of human behavior. Mandatory observations, observation quotas, names on observation cards, disciplinary action, employee blame—certainly not the makings of a behavioral safety process. One of the folks Reggie spoke to was adamant in sharing how they felt about their BBS program: "BBS is B.S.! BBS is trying to blame us for all the incidents. Why would you support that, Reggie?" One of the plant managers was also fed up with their BBS program, arguing: "We invest tens of thousands of dollars a year on BBS and we keep seeing the same three at-risk behaviors show up over and over. It doesn't work." Reggie was put on the spot. There were clearly misconceptions about behavioral safety and behavioral science. He knew he had to defend his behavioral program, but also had to make some clarifications. Where to begin?

As we wrote this book, we deliberated on whether to include a section on criticisms in the market levied at behavioral safety. On the one hand, we hope that by reading this book about the science behind behavioral safety, the misconceptions will be readily apparent and easily dismissible. On the other hand, by not acknowledging the misconceptions, we would not want to be perceived as denying their existence (the misconceptions are out there) or as giving truth to them.[1] We decided it best to describe the common misconceptions and explain why they are inaccurate descriptions of a behavioral safety program based on behavioral science.

BEHAVIORAL SAFETY BLAMES THE WORKER

Perhaps the most common misconception is that behavioral safety blames the worker for their behavior. A case study was presented in which the San Francisco Bay Bridge Company suppressed the reporting of injuries and near misses (Brown & Barab, 2007). The authors claimed, without any data, that this was due to the company's behavioral safety program:

DOI: 10.4324/9781003290711-9

> BBS is a safety management system based on the assumption that most injuries are caused by employees' "unsafe" behaviors. According to this theory, "correcting" these behaviors through positive or negative incentives – instead of identifying and eliminating physical and systemic workplace hazards – can modify workers' behavior, resulting in improved work site safety and significantly reduced injury and illness rates (Brown & Barab, 2007, p. 312).

Instead, the real reason for the suppressed reporting turned out to be a form of dysfunctional behavior management, including:

> 1) cash incentives to workers and supervisors who do not report injuries; 2) reprisals and threats of reprisals against those employees who do report injuries; 3) selection and use of employer friendly occupational health clinics and workers compensation insurance administrators; 4) strict limits on the activities of contract industrial hygiene consultants; and 5) a secretive management committee that decides whether reported injuries and illnesses are legitimate and recordable (Brown & Barab, 2007, p. 313).

Proponents of this position claim that behavioral safety blames the worker by focusing on their behavior and not the systematic causes of risk and injury. But this perspective is false and Skinner himself would tell you that. He famously recounted a time when his rats were not behaving the way he had hoped and he found himself blaming the rats. Then he realized that *the organism is always right*! By that he meant that he, as the experimenter, had set up the environment for the rats, and the rats were just behaving the way the contingencies dictated. Similarly, behavioral safety focuses on identifying the environmental variables, including hazards, that influence behavior (see Science Moment 1.5 for how the environment selects behavior and Chapter 6 on trending and functional analysis for a more in-depth discussion of environmental influences).

Human behavior interacting with a hazard is the last thing that happens before most injury. Therefore, an uninformed analysis may suggest the worker is to blame. However, as we have discussed, there is a complex (and sometimes not too complex) interplay of antecedents, consequences and interlocks leading up to the very moment the at-risk behavior occurs. Our role in behavioral safety is to identify the most critical contingencies that lead to at-risk behavior and then change the contingencies so the safe behavior can emerge. Take the person out of the equation and you arrive at our scientific perspective. For further discussion of this point, read Ludwig's (2018) book *Dysfunctional Practices that Kill Your Safety Culture (And What to Do about Them)*.

BEHAVIORAL SAFETY LACKS EMPIRICAL EVIDENCE

Critics have claimed that BBS is built on flawed research or has no empirical backing. Frederick and Lessin (2000) and Howe (2001) state that the empirical evidence of BBS is based on Heinrich's work (1931), with no mention of Skinner or any other modern behavioral researchers discussed throughout this book:

> *Unfortunately, behavior-based safety programs are just a retread of old outdated ideas and strategies that have never proven effective* (Howe, 2001, p. 6).
>
> *Behavioral safety claims to be based on science. Its advocates are somewhat entitled to the claim. The strength of the effect is not well documented. Reproducibility is one issue. It is nearly impossible for even the best intentioned researchers to find or create situations that are enough alike for evaluation and to include a control group. A meta-analysis of behavior-based safety (BBS) interventions concludes: "A statistically significant reduction in injuries/incidents was observed after conducting a BBS intervention in a workplace." However, this statistical significance should be interpreted with caution, due to the methodological quality of studies included in the meta-analysis. Reliable results require interventions with high methodological quality based on the specific needs of the workplace* (Metzgar, 2011, p. 20).
>
> *BBS is largely based on experiments with rodents"* and *"the very foundation of BBS is probably based on inapplicable science* (Eckenfelder, 2003).

The misconception that behavioral safety has no empirical backing or is built on flawed research could not be further from the truth. As we described back in Science Moment 1.1, behavioral safety is built on the foundational principles of behavior discovered in behavior analysis. Those principles have been replicated both inside labs and in real-world settings, with non-humans and humans alike, for over 100 years. Behavioral science works. Behavioral safety specifically has a long history of successfully demonstrating effective results and has been replicated across many industries and organizations (see Chapter 8 on evaluation, where we highlighted dozens of successful applications. See also Alavosius & Burleigh (2022) for a chronological list of over 300 research articles, texts, and conference proceedings highlighting the history of behavioral safety).

THE LAST WORD IS DATA

Behavior analysis and its applications to behavioral safety have a long history of success. Beyond empirical research, however, it is the work of

organizations—as exemplified by the Cambridge Center for Behavioral Studies (CCBS) accredited programs—that have consistently demonstrated the
effectiveness of our science in workplace safety. Figure 9.1 highlights the value
of behavioral safety across all CCBS accredited programs that provided sufficient data. In this graph, we compare the average and individual Occupational
Health and Safety Administration recordable incident rate (ORIR) for CCBS
accredited programs to average ORIR across all U.S. industries (BLS, 2021).
Each program's data is presented in light gray behind the average injury rate
for all CCBS accredited programs. Regardless of industry, the CCBS programs
with behavioral safety programs consistently exhibit lower ORIR than their
peers. These data provide our final evidence for the effectiveness of behavioral
safety in the reduction of injuries.

What can be lost from our scientific explanations and data is the impact
that behavioral safety has on the lives and wellbeing of individuals outside
the workplace. If workers suffer a life-altering injury, they have less access to

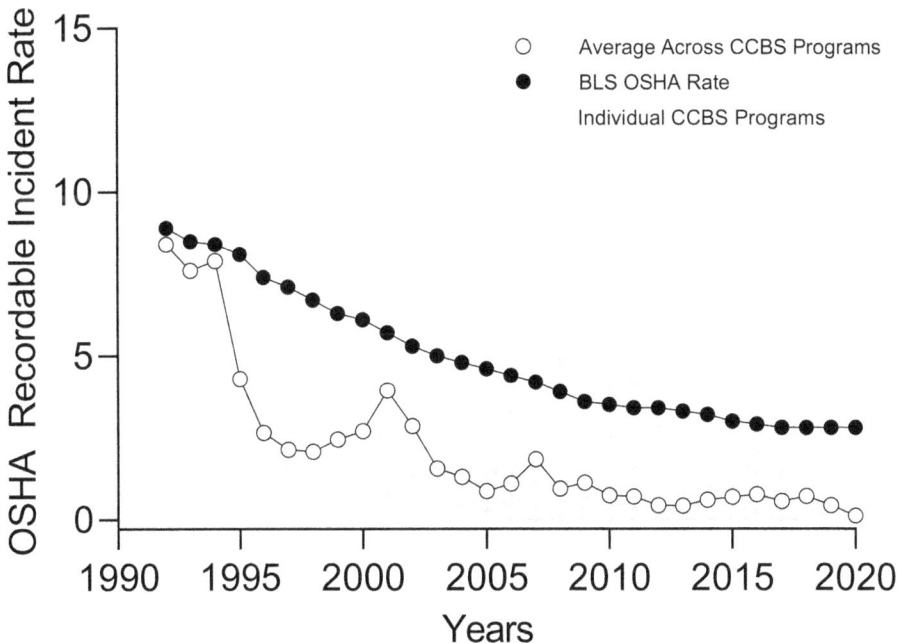

Figure 9.1   All Accredited Programs OSHA Rate Compared to BLS Data

ORIR on the y-axis and consecutive years on the x-axis. Black circles indicate injury rates per 100
full-time equivalent (FTE) workers across all U.S. industries (BLS, 2021). White circles indicate
injury rates per 100 FTE workers averaged across CCBS accredited sites after implementing a
behavioral safety program. Gray circles indicate individual accredited program ORIRs. Data from
the accredited programs were used with permission from the CCBS. From "Behavioral Safety:
An Efficacious Application of Applied Behavior Analysis to Reduce Human Suffering," by T.D.
Ludwig and M.M. Laske, *Journal of Organizational Behavior Management* (Taylor & Francis, 2022).

reinforcers in their daily life activities, which may lead to lower satisfaction and enjoyment for them and their family. Behavioral safety programs (i.e., observation, feedback, analysis, intervention, evaluation) shape safe behaviors and build a degree of mastery across work tasks that leads to the adoption of new safety behaviors. We strongly suspect that participation in these programs also increases the likelihood that workers will engage in safe behaviors outside of work. It is likely that participants will use their behavioral skills to impact the safety of their families and communities. Therefore, we maintain that behavioral safety influences socially significant behaviors, which has implications beyond the workplace. Behavioral safety is thus yet another efficacious application of behavior analysis for the improvement of the human condition.

The advancement of our science is not complete, however. We conclude with a challenge for the field. For behavioral safety to continue contributing to socially significant events, we must continue to target workplace fatalities (permanent removal of access to reinforcers); process safety events (which impact people and our environment); and address other emerging threats in an increasingly technological and global workplace. We are hopeful for the future and encourage the next wave of behavior analysts and practitioners to target these pertinent issues so behavior analysis can continue to contribute to the wellbeing of humankind.

Note

1. Although we will be dispelling the myths about behavioral safety, there are certainly "BBS" programs that are not based on the science of behavior. Therefore, we believe are certainly "truths" to these perceptions in relation to BBS programs not based on the science and practices we described in this book.

REFERENCES

Agnew, J.L., & Daniels, A.C. (2010). *Safe by accident.* Performance Management Publications.

Agnew, J.L., & Redmon, W.K. (1993). Contingency specifying stimuli: The role of "rules" in organizational behavior management. *Journal of Organizational Behavior Management*, 12(2), 67–76. https://doi.org/10.1300/J075v12n02_04

Agnew, J.L., & Snyder, G. (2008). *Removing obstacles to safety a behavior-based approach.* Performance Management Publications.

Agnew, J.L., Uhl, D. (2021). *Safe by design: A behavioral systems approach to human performance improvement.* Performance Management Publications.

Ajayi, A., Oyedele, L., Akinade, O., Bilal, M., Owolabi, H., Akanbi, L., & Delgado, J.M.D. (2020). Optimised Big Data analytics for health and safety hazards prediction in power infrastructure operations. *Safety Science*, 125. https://doi.org/10.1016/j.ssci.2020.104656

Alavosius, M.P., & Burleigh, K. (2022). Behavior-based safety as a replicable technology. In R.A. Houmanfar, M. Fryling, & M.P. Alavosius (eds.), *Applied behavior science in organizations: Consilience of historical and emerging trends in organizational behavior management* (pp. 21–63). Routledge. https://doi.org/10.4324/9781003198949-2

Alvero, A.M., & Austin, J. (2004). The effects of conducting behavioral observations on the behavior of the observer. *Journal of Applied Behavior Analysis*, 37(4), 457–468. https://doi.org/10.1901/jaba.2004.37-457

Alvero, A.M., Bucklin, B.R., & Austin, J. (2001). An objective review of the effectiveness and essential characteristics of performance feedback in organizational settings (1985–1998). *Journal of Organizational Behavior Management*, 21(1), 3–29. https://doi.org/10.1300/J075v21n01_02

Alvero, A.M., Rost, K., & Austin, J. (2008). The safety observer effect: The effects of conducting safety observations. *Journal of Safety Research*, 39(4), 365–373. https://doi.org/10.1016/j.jsr.2008.05.004

Austin, J. (2000). Performance analysis and performance diagnostics. In J. Austin & J.E. Carr (eds.), *Handbook of applied behavior analysis* (pp. 321–349). Context Press.

Austin, J., Kessler, M.L., Riccobono, J.E., & Bailey, J.S. (1996). Using feedback and reinforcement to improve the performance and safety of a roofing crew. *Journal of Organizational Behavior Management*, 16(2), 49–75. https://doi.org/10.1300/J075v16n02_04

Baer, D.M., Peterson, R.F., & Sherman, J.A. (1967). The development of imitation by reinforcing behavioral similarity to a model. *Journal of the Experimental Analysis of Behavior*, 10(5), 405–416. https://doi.org/10.1901/jeab.1967.10-405

Balcazar, F.E., Hopkins, B.L., & Suarez, Y. (1985). A critical, objective review of performance feedback. *Journal of Organizational Behavior Management*, 7(3–4), 65–89. https://doi.org/10.1300/J075v07n03_05

Bateman, M.J., & Ludwig, T.D. (2003). Managing distribution quality through an adapted incentive program with tiered goals and feedback. *Journal of Organizational Behavior Management, 23*(1), 33–55. https://doi.org/10.1300/J075v23n01_03

Berger, S.M., & Ludwig, T.D. (2007). Reducing warehouse employee errors using voice-assisted technology that provided immediate feedback. *Journal of Organizational Behavior Management, 27*(1), 1–31. https://doi.org/10.1300/J075v27n01_01

Berglund, K.M., & Ludwig, T.D. (2009). Approaching error-free customer satisfaction through process change and feedback systems. *Journal of Organizational Behavior Management, 29*(1), 19–46. https://doi.org/10.1080/01608060802660140

Binder, C. (1996). Behavioral fluency: Evolution of a new paradigm. *Behavior Analyst, 19*(2), 163–197. https://doi.org/10.1007/BF03393163

Binder, C. (1999). Fluency development. In D.G. Langdon, K.S. Whiteside, & M.M. McKenna (eds.), *Intervention resource guide: 50 performance improvement tools* (pp. 176–183). Jossey-Bass.

Binder, C. (2016). Integrating organizational-cultural values with performance management. *Journal of Organizational Behavior Management, 36*(2–3), 185–201. https://doi.org/10.1080/01608061.2016.1200512

Binder, C. (2017). What it really means to be accomplishment based. *Performance Improvement, 56*(4), 20–25. https://doi.org/10.1002/pfi

Binder, C. (2022). From fluency-based instruction to accomplishment-based performance improvement. In R.A. Houmanfar, M. Fryling, & M.P. Alavosius (eds.), *Applied behavior science in organizations: Consilience of historical and emerging trends in organizational behavior management* (pp. 81–98). Routledge.

Binder, C., & Sweeney, L. (2002). Building fluent performance in a customer call center. *Performance Improvement, 41*(2), 29–37. https://doi.org/10.1002/pfi.4140410207

Blackman, A.L., Novak, M.D., DiGennaro Reed, F.D., & Erath, T.G. (2022). The efficacy of variations of observation and data recording on trainee performance. *Journal of Organizational Behavior Management.* Advance online publication. https://doi.org/10.1080/01608061.2021.1979708

Blakely, E., & Schlinger, H. (1987). Rules: Function-altering contingency-specifying stimuli. *The Behavior Analyst, 10*(2), 183–187.

Blasingame, A., Hale, S., & Ludwig, T.D. (2014). The effects of employee-led process design on welder set-up intervals. *Journal of Organizational Behavior Management, 34*(3), 207–222. https://doi.org/10.1080/01608061.2014.944745

Bogard, K., Ludwig, T.D., Staats, C., & Kretschmer, D. (2015). An industry's call to understand the contingencies involved in process safety: Normalization of deviance. *Journal of Organizational Behavior Management, 35*(1–2), 70–80. https://doi.org/10.1080/01608061.2015.1031429

Boiral, O., Guillaumie, L., Heras-Saizarbitoria, I., & Tayo Tene, C.V. (2018). Adoption and outcomes of ISO 14001: A systematic review. *International Journal of Management Reviews, 20*(2), 411–432. https://doi.org/10.1111/ijmr.12139

Boyce, T.E., & Geller, E.S. (2001). Applied behavior analysis and occupational safety. *Journal of Organizational Behavior Management, 21*(1), 31–60. https://doi.org/10.1300/J075v21n01_03

Brache, A.P., & Rummler, G.A. (1997). Managing an organization as a system. *Training*.

Brethower, D.M. (1972). *Behavior analysis in business and industry: A total performance system*. Behaviordelia.

Brethower, D.M. (1982). The total performance system. In R.M. O'Brien, A.M. Dickinson, & M.P. Rosow (eds.), *Industrial behavior modification: A management handbook* (pp. 250–369). Pergamon Press.

Brethower, D.M., & Dams, P. (1999). Systems thinking (and systems doing). *Performance Improvement*, *38*(1), 37–52.

Bureau of Labor Statistics [BLS]. (2019a). How to compute a firm's incidence rate for safety management. U.S. Department of Labor. https://www.bls.gov/iif/osheval.htm

Bureau of Labor Statistics [BLS]. (2019b). 2018 survey of occupational injuries & illnesses: Charts package. U.S. Department of Labor. https://stats.bls.gov/iif/soii-charts-2018.pdf

Bureau of Labor Statistics [BLS]. (2020a). Employer-reported workplace injuries and illnesses–2019 (USDL-20–2030). U.S. Department of Labor. https://www.bls.gov/news.release/pdf/osh.pdf

Bureau of Labor Statistics [BLS]. (2020b). National census of fatal occupational injuries in 2019 (USDL-20–2265). U.S. Department of Labor. https://www.bls.gov/news.release/pdf/cfoi.pdf

Bureau of Labor Statistics [BLS]. (2021). Survey of occupational injuries and illnesses data. U.S. Department of Labor. https://www.bls.gov/iif/soii-data.htm

Bumstead, A., & Boyce, T.E. (2005). Exploring the effects of cultural variables in the implementation of behavior-based safety in two organizations. *Journal of Organizational Behavior Management*, *24*(4), 43–63. https://doi.org/10.1300/J075v24n04_03

Cambridge Center for Behavioral Studies (2005). *Accreditation application: Halliburton Gulf of Mexico*. https://behavior.org/wp-content/uploads/2017/06/289.pdf

Cambridge Center for Behavioral Studies (2006). *PBBS reaccreditation application: Eastman Chemical Acetate Fibers Division*. https://behavior.org/wp-content/uploads/2017/06/279.pdf

Cambridge Center for Behavioral Studies (2009). *Application for accreditation of a behavior based safety program that replicates an accredited program*. https://behavior.org/wp-content/uploads/2017/06/291.pdf

Cambridge Center for Behavioral Studies (2012a). *Application for re-accreditation of a behavior based safety program: Advantage Logistics Southeast Regional Facility (SERF)*. https://behavior.org/help-centers/safety/Accredited%20Companies/

Cambridge Center for Behavioral Studies (2012b). *Application for re-accreditation of safety programs based on the principles of behavior: Marathon Petroleum Company LLC, Illinois Refining Division*. https://behavior.org/wp-content/uploads/2017/06/839.pdf

Cambridge Center for Behavioral Studies (2012c). *SENCO Construction Inc. Application for the accreditation of safety programs on the principles of behavior*. https://behavior.org/wp-content/uploads/2017/06/837.pdf

Cambridge Center for Behavioral Studies (2012d). *Gribbins Insulation CCBS application.* https://behavior.org/wp-content/uploads/2017/06/841.pdf

Cambridge Center for Behavioral Studies (2013). *Application for re-accreditation of a behavior based safety program: Advantage Logistics, MRDC.*

Cambridge Center for Behavioral Studies (2014a). *Application for the accreditation of safety programs on the principles of behavior: Marathon Petroleum Company, Michigan Refining Division.* https://behavior.org/wp-content/uploads/2017/06/835.pdf

Cambridge Center for Behavioral Studies (2014b). *Written application for the reaccreditation of the Costain Behavioural Safety Programme by the Cambridge Center for Behavioral Studies.* https://behavior.org/wp-content/uploads/2021/03/CBS_Reaccreditation_Submission_Final.pdf

Cambridge Center for Behavioral Studies (2015a). *Acetate Fibers Division: CCBS accreditation application.* https://behavior.org/wp-content/uploads/2017/06/896.pdf

Cambridge Center for Behavioral Studies (2015b). *SENCO Construction Inc. application for the accreditation of safety programs on the principles of behavior.* https://behavior.org/wp-content/uploads/2017/06/942.pdf

Cambridge Center for Behavioral Studies (2015c). *SDR Coating Company: Application for accreditation.* https://behavior.org/wp-content/uploads/2017/06/951.pdf

Cambridge Center for Behavioral Studies (2015d). Western Energy Company: Application for accreditation of safety program based on the principles of behavior. https://behavior.org/wp-content/uploads/2017/06/940.pdf

Cambridge Center for Behavioral Studies (2016). *Application for the accreditation of safety processes on the principles of behavior: Marathon Petroleum Company, LP, Ohio Refining Division.* https://behavior.org/wp-content/uploads/2017/06/934.pdf

Cambridge Center for Behavioral Studies (2017a). *Application for the re-accreditation of safety processes on the principles of behavior: Marathon Petroleum Company, LP, Texas Refining Division.* https://behavior.org/wp-content/uploads/2018/02/2017MarathonTRDRecertificationApplication.pdf

Cambridge Center for Behavioral Studies (2017b). *Application for the re-accreditation of safety programs on the principles of behavior: Marathon Petroleum Company Michigan Refining Division.* https://behavior.org/wp-content/uploads/2018/02/MRDReAccreditationApplication2017.pdf

Cambridge Center for Behavioral Studies (2018a). *Ahlstrom-Munksjo Italia: Cambridge Center for Behavioral Studies accreditation site visit report.* https://behavior.org/wp-content/uploads/2019/10/AhlstromMunksjo2018.pdf

Cambridge Center for Behavioral Studies (2018b). *Behavioral safety program description and accreditation application: Costain Limited.* https://behavior.org/wp-content/uploads/2018/08/COSTAIN-BSA-2018Application.pdf

Cambridge Center for Behavioral Studies (2018c, October 10). *Behavioral safety program description and accreditation application: SENCO Construction, Inc.* https://behavior.org/wp-content/uploads/2020/03/SENCO-Application.pdf

Cambridge Center for Behavioral Studies (2019). *Behavioral safety program description and application: Marathon Petroleum Corporation FUELS (Forever Uniting Employees Lives through Safety).* https://behavior.org/wp-content/uploads/2020/03/Marathon-IRD-Application-with-Observation-Attachment.pdf

Cambridge Center for Behavioral Studies (2020a). *Program description: MRD Circle of Safety Program Marathon Petroleum Company.* https://behavior.org/wp-content/uploads/2021/03/MRD-2020-Program-Description-Final.pdf

Cambridge Center for Behavioral Studies (2020b). *Program description: St. Paul Park AWARE program Marathon Petroleum Company.* https://behavior.org/wp-content/uploads/2021/01/AWARE-StPaulPark-Program-Application.pdf

Cambridge Center for Behavioral Studies (2022a). *CCBS behavioral safety accreditation standards.* https://behavior.org/help-centers/safety/

Cambridge Center for Behavioral Studies (2022b). *Companies achieving behavioral accreditation.* https://behavior.org/help-centers/safety/Accredited%20Companies/

Camden, M.C., Price, V.A., & Ludwig, T.D. (2011). Reducing absenteeism and rescheduling among grocery store employees with point-contingent rewards. *Journal of Organizational Behavior Management, 31*(2), 140–149. https://doi.org/10.1080/01608061.2011.569194

Campbell, D.T., & Stanley, J.C. (1963). *Experimental and quasi-experimental designs for research.* Rand McNally.

Catania, A.C. (1979). *Learning.* Prentice-Hall.

Catania, A.C. (1991). The gifts of culture and of eloquence: An open letter to Michael J. Mahoney in reply to his article, "Scientific psychology and radical behaviorism." *The Behavior Analyst, 14*(1), 61–72. https://doi.org/10.1007/bf03392553

Cervone, D., Jiwani, N., & Wood, R. (1991). Goal setting and the differential influence of self-regulatory processes on complex decision-making performance. *Journal of Personality and Social Psychology, 61*(2), 257–266. https://doi.org/10.1037/0022-3514.61.2.257

Chance, P. (2006). *Learning & behavior* (5th ed.). Thomson Wadsworth.

Chi, C.F., Lin, S.Z., & Dewi, R.S. (2014). Graphical fault tree analysis for fatal falls in the construction industry. *Accident Analysis and Prevention, 72*, 359–369. https://doi.org/10.1016/j.aap.2014.07.019

Chiang, Y.H., Wong, F.K.W., & Liang, S. (2018). Fatal construction accidents in Hong Kong. *Journal of Construction Engineering and Management, 144*(3), 1–11. https://doi.org/10.1061/(ASCE)CO.1943-7862.0001433

Cooper, M.D. (2006). Exploratory analyses of the effects of managerial support and feedback consequences on behavioral safety maintenance. *Journal of Organizational Behavior Management, 26*(3), 1–41. https://doi.org/10.1300/J075v26n03_01

Cooper, M.D. (2009). Behavioral safety interventions: A review of process design factors. *Professional Safety, 54*(2), 36.

Cooper, M.D. (2019). The efficacy of industrial safety science constructs for addressing serious injuries & fatalities (SIFs). *Safety Science, 120*, 164–178. https://doi.org/10.1016/j.ssci.2019.06.038

Cooper, M.D., Phillips, R.A., Sutherland, V.J., & Makin, P.J. (1994). Reducing accidents using goal setting and feedback: A field study. *Journal of Occupational and Organizational Psychology, 67*(3), 219–240. https://doi.org/10.1111/j.2044-8325.1994.tb00564.x

Dagen, J.C., Alavosius, M.P., & Harshbarger, D. (2009). *Using managerial feedback to improve safety in the petroleum industry* [Unpublished manuscript]. University of Nevada, Reno, Reno, NV.

Daniels, A.C. (1989). *Performance management* (3rd ed.). Performance Management Publications.

Daniels, A.C., & Bailey, J.S. (2014). *Performance management changing behavior that drives organizational effectiveness* (5th ed.). Aubrey Daniels International, Inc.

DeJoy, D.M. (2005). Behavior change versus culture change: Divergent approaches to managing workplace safety. *Safety Science, 43*(2), 105–129. https://doi.org/10.1016/j.ssci.2005.02.001

Dembe, A.E. (2001). The social consequences of occupational injuries and illnesses. *American Journal of Industrial Medicine, 40*(4), 403–417. https://doi.org/10.1002/ajim.1113

Deming, E.W. (1982). *Quality, productivity, and competitive position.* Center for Advanced Engineering Study, MIT.

Deming, E.W. (1986). *Out of the crisis.* Center for Advanced Engineering Study, MIT.

DePaolo, J., Gravina, N.E., & Harvey, C. (2019). Using a behavioral intervention to improve performance of a women's college lacrosse team. *Behavior Analysis in Practice, 12*(2), 407–411. https://doi.org/10.1007/s40617-018-0272-6

Depasquale, J.P., & Geller, E.S. (1999). Critical success factors for behavior-based safety: A study of twenty industry-wide applications. *Journal of Safety Research, 30*(4), 237–249. https://doi.org/10.1016/S0022-4375(99)00019-5

DeRiso, A., & Ludwig, T.D. (2012). An investigation of response generalization across cleaning and restocking behaviors in the context of performance feedback. *Journal of Organizational Behavior Management, 32*(2), 140–151. https://doi.org/10.1080/01608061.2012.676500

Dickinson, A.M. (1989). The detrimental effects of extrinsic reinforcement on "intrinsic motivation." *The Behavior Analyst, 12*(1), 1–15. https://doi.org/10.1007/BF03392473

Diener, L.H., McGee, H.M., & Miguel, C.F. (2009). An integrated approach for conducting a behavioral systems analysis. *Journal of Organizational Behavior Management, 29*(2), 108–135. https://doi.org/10.1080/01608060902874534

DiGennaro Reed, F.D., Novak, M.D., Erath, T.G., Brand, D., & Henley, A.J. (2018). Pinpointing and measuring employee behavior. *Organizational Behavior Management: The Essentials,* 143–168.

Doll, J., Livesey, J., McHaffie, E., & Ludwig, T.D. (2007). Keeping an uphill edge: Managing cleaning behaviors at a ski shop. *Journal of Organizational Behavior Management, 27*(3), 41–60. https://doi.org/10.1300/J075v27n03_04

Ehrlich, R.J., Nosik, M.R., Carr, J.E., & Wine, B. (2020). Teaching employees how to receive feedback: A preliminary investigation. *Journal of Organizational Behavior Management, 40*(1–2), 19–29. https://doi.org/10.1080/01608061.2020.1746470

Eikenhout, N., & Austin, J. (2004). Using goals, feedback, reinforcement, and a performance matrix to improve customer service in a large department store. *Journal of Organizational Behavior Management, 24*(3), 27–62. https://doi.org/10.1300/J075v24n03_02

Ezerins, M.E., Laske, M.M., Hinson, P., & O'Neil, T. (2020). *Organizational development report: Cross-functional data integration into safety, production, and maintenance* [Unpublished manuscript]. Department of Psychology, Appalachian State University.

Ezerins, M.E., Ludwig, T.D., O'Neil, T., Foreman, A.M., & Açıkgöz, Y. (2022). Advancing safety analytics: A diagnostic framework for assessing system readiness within occupational safety and health. *Safety Science, 146*. https://doi.org/10.1016/j.ssci.2021.105569

Fante, R., Gravina, N., Betz, A., & Austin, J. (2010). Structural and treatment analyses of safe and at-risk behaviors and postures performed by pharmacy employees. *Journal of Organizational Behavior Management, 30*(4), 325–338. https://doi.org/10.1080/01608061.2010.520143

Fellner, D.J., & Sulzer-Azaroff, B. (1984). Increasing industrial safety practices and conditions through posted feedback. *Journal of Safety Research, 15*(1), 7–21. https://doi.org/10.1016/0022-4375(84)90026-4

Ferster, C.B., & Skinner, B.F. (1957). *Schedules of reinforcement.* Prentice-Hall.

Fox, C.J., & Sulzer-Azaroff, B. (1989). The effectiveness of two different sources of feedback on staff teaching of fire evacuation skills. *Journal of Organizational Behavior Management, 10*(2), 19–35. https://doi.org/10.1300/J075v10n02_03

Frederick, J., & Lessin, N. (2000). Blame the worker – The rise of behavioral-based safety programs. *Multinational Monitor.*

Geller, E.S. (1984). A delayed reward strategy for large-scale motivation of safety belt use: A test of long-term impact. *Accident Analysis and Prevention, 16*(5–6), 457–463. https://doi.org/10.1016/0001-4575(84)90058-7

Geller, E.S. (1996). *The psychology of safety.* Chilton Book Company.

Geller, E.S. (2001a). *Keys to behavior-based safety.* Government Institutes.

Geller, E.S. (2001b). *The psychology of safety* (2nd ed.). CRC Press.

Geller, E.S. (2002a). Building a belief in safety. *Industrial Safety & Hygiene News, 36*(4), 17–18.

Geller, E. S. (2002b). Fueling the participation factor. *Industrial Safety & Hygiene News, 36*(2), 17–18.

Geller, E S. (2002c). *The participation factor: How to increase involvement in occupational safety.* American Society of Safety Engineers.

Geller, E.S. (2005a). Behavior-based safety and occupational risk management. *Behavior Modification, 29*(3), 539–561. https://doi.org/10.1177/0145445504273287

Geller, E.S. (2005b). *People-based safety: The source.* Coastal Training Technologies Corporation.

Geller, E.S., Berry, T.D., Ludwig, T.D., Evans, R.E., Gilmore, M.R., & Clarke, S.W. (1990). A conceptual framework for developing and evaluating behavior change interventions for injury control. *Health Education Research: Theory and Practice, 5*(2), 125–137. https://doi.org/10.1093/her/5.2.125

Geller, E.S., Kalsher, M.J., Rudd, J.R., & Lehman, G.R. (1989). Promoting safety belt use on a university campus: An integration of commitment and incentive strategies. *Journal of Applied Social Psychology, 19*(1), 3–19. https://doi.org/10.1111/j.1559-1816.1989.tb01217.x

Geller E.S., & Ludwig, T.D. (1991). A behavior change taxonomy for improving road safety. In M.J. Koornstra & J. Christensen (eds.) *Enforcement and rewarding: Strategies and effects*, pp. 41–45. Proceedings of the Organization for Economic Co-operation and Development International Road Safety Conference, Copenhagen, Denmark.

Giacalone, R.A., & Rosenfeld, P. (1986). Self-presentation and self-promotion in an organizational setting. *Journal of Social Psychology, 126*(3), 321–326. https://doi.org/10.1080/00224545.1986.9713592

Gilbert, T.F. (1978). *Human competence: Engineering worthy performance.* McGraw-Hill.

Goomas, D.T., & Ludwig, T.D. (2007). Enhancing incentive programs with proximal goals and immediate feedback: Engineered labor standards and technology enhancements in stocker replenishment. *Journal of Organizational Behavior Management, 27*(1), 33–68. https://doi.org/10.1300/J075v27n01

Goomas, D.T., & Ludwig, T.D. (2009). Standardized goals and performance feedback aggregated beyond the work unit: Optimizing the use of engineered labor standards and electronic performance monitoring. *Journal of Applied Social Psychology, 39*(10), 2425–2437. https://doi.org/10.1111/j.1559-1816.2009.00532.x

Goomas, D.T., Smith, S.M., & Ludwig, T.D. (2011). Business activity monitoring: Real-time group goals and feedback using an overhead scoreboard in a distribution center. *Journal of Organizational Behavior Management, 31*(3), 196–209. https://doi.org/10.1080/01608061.2011.589715

Gravina, N., Austin, J., & Kazbour, R. (2014, June 8–11). *Taking incident investigations to the next level: A behavioral science approach* [Paper Presentation]. ASSE Professional Development Conference and Exposition, Orlando, Florida, United States.

Gravina, N., Cummins, B., & Austin, J. (2017). Leadership's role in process safety: An understanding of behavioral science among managers and executives is needed. *Journal of Organizational Behavior Management, 37*(3–4), 316–331. https://doi.org/10.1080/01608061.2017.1340925

Greer, R.D., Dudek-Singer, J., & Gautreaux, G. (2006). Observational learning. *International Journal of Psychology, 41*(6), 486–499. https://doi.org/10.1080/00207590500492435

Grindle, A.C., Dickinson, A.M., & Boettcher, W. (2000). Behavioral safety research in manufacturing settings: A review of the literature. *Journal of Organizational Behavior Management, 20*(1), 29–68. https://doi.org/10.1300/J075v20n01_03

Guo, B.H.W., Goh, Y.M., & Le Xin Wong, K. (2018). A system dynamics view of a behavior-based safety program in the construction industry. *Safety Science, 104*, 202–215. https://doi.org/10.1016/j.ssci.2018.01.014

Hagge, M., McGee, H., Matthews, G., & Aberle, S. (2017). Behavior-based safety in a coal mine: The relationship between observations, participation, and injuries over a 14-year period. *Journal of Organizational Behavior Management, 37*(1), 107–118. https://doi.org/10.1080/01608061.2016.1236058

Heinrich, H.W. (1931). *Industrial accident prevention: A scientific approach*. McGraw-Hill.

Hensler, D.R., Marquis, M.S., Abrahamse, A.F., Berry, S.H., Ebener, P.A., Lewis, E.G., Lind, E. A., MacCoun, R.J., Manning, W.G., Rogowski, J.A., & Vaiana, M.E. (1991). *Compensation for accidental injuries in the United States*. RAND. https://www.rand.org/pubs/reports/R3999.html

Heras-Saizarbitoria, I., & Boiral, O. (2013). ISO 9001 and ISO 14001: Towards a research agenda on management system standards. *International Journal of Management Reviews*, 15(1), 47–65. https://doi.org/10.1111/j.1468-2370.2012.00334.x

Hermann, J.A., Ibarra, G.V., & Hopkins, B.L. (2010). A safety program that integrated behavior-based safety and traditional safety methods and its effects on injury rates of manufacturing workers. *Journal of Organizational Behavior Management*, 30(1), 6–25. https://doi.org/10.1080/01608060903472445

Hickman, J.S., & Geller, E.S. (2003). A safety self-management intervention for mining operations. *Journal of Safety Research*, 34(3), 299–308. https://doi.org/10.1016/S0022-4375(03)00032-X

Hinson, P.E. & Laske, M.M. (2020). *Analysis of scheduling & weather variables and incidents/near misses reported* [Unpublished manuscript]. Department of Psychology, Appalachian State University.

Holweg, M. (2007). The genealogy of lean production. *Journal of Operations Management*, 25(2), 420–437. https://doi.org/10.1016/j.jom.2006.04.001

Howe, J. (2001). Warning! Behavior-based safety can be hazardous to your health and safety program! A union critique of behavior-based safety. UAW, International Union. http://www.uawlocal974.org/BSSafety/Warning!_Behavior-Based_Safety_Can_Be_Hazardous_To_Your_Health_and_Safety_Program!.pdf

Huang, L., Wu, C., Wang, B., & Ouyang, Q. (2018). Big-data-driven safety decision-making: A conceptual framework and its in fl uencing factors. *Safety Science*, 109, 46–56. https://doi.org/10.1016/j.ssci.2018.05.012

Hyten, C. (2009). Strengthening the focus on business results: The need for systems approaches in organizational behavior management. *Journal of Organizational Behavior Management*, 29(2), 87–107. https://doi.org/10.1080/01608060902874526

Hyten, C., & Ludwig, T.D. (2017). Complacency in process safety: A behavior analysis toward prevention strategies. *Journal of Organizational Behavior Management*, 37(3–4), 240–260. https://doi.org/10.1080/01608061.2017.1341860

Hyten, C., Ludwig, T.D., Moran, D.J., McSween, T., & Grindle, A. (2017). Worker behavior: Constructing a straw man argument. *Professional Safety*, July, 9.

Imai, M. (1986). *Kaizen: The key to Japan's competitive success*. McGraw-Hill.

Johnson, C.M., Mawhinney, T.C., & Redmon, W.K. (2001). *Handbook of organizational performance: Behavior analysis & management*. Haworth Press, Inc.

Johnston, J.M., & Pennypacker, H.S. (1980). *Strategies and tactics of human behavioral research*. Lawrence Erlbaum Associates, Inc.

Kazdin, A.E. (1994). *Behavior Modification in Applied Settings* (4th ed.). Brookes/Cole Publishing Company.

Keller, F.S., & Schoenfeld, W.N. (1950). *Principles of psychology*. Appleton-Century-Crofts.

King, A., Gravina, N., & Sleiman, A. (2018). Observing the observer. *Journal of Organizational Behavior Management, 38*(4), 1–18. https://doi.org/10.1080/01608061.2018.1514346

Komaki, J.L., Barwick, K.D., & Scott, L.R. (1978). A behavioral approach to occupational safety: Pinpointing and reinforcing safe performance in a food manufacturing plant. *Journal of Applied Psychology, 63*(4), 434–445. https://doi.org/10.1037/0021-9010.63.4.434

Krause, T.R. (1997). *The behavior-based safety process: Managing involvement for an injury-free culture* (2nd ed.). John Wiley & Sons, Inc.

Kretschmer, D.C. (2015). *Assessing the efficacy of training targeting contextual comments in behavior based safety observations* [Unpublished master's thesis]. Appalachian State University.

Laske, M.M., Açıkgöz, Y., Ludwig, T.D., & Bergman, S. (2020, May 21–25). Utilizing data analytics to inform safety interventions and reduce adverse safety outcomes. In J.E. Friedel (Chair) *Big data and analytics in behavior analysis* [Symposium]. Association for Behavior Analysis International 46th Annual Conference, Washington DC, United States.

Laske, M.M., Hinson, P.E., Açıkgöz, Y., Ludwig, T.D., Foreman, A.M., & Bergman, S.M. (2022). Do employees' work schedules put them at-risk? The role of shift scheduling and holidays in predicting near miss and incident likelihood. *Journal of Safety Research*. Advance online publication. https://doi.org/10.1016/j.jsr.2022.07.015

Laske, M.M., & Ludwig, T.D. (2022a). *Pinpointing counts! A proposed criteria for evaluating discriminant pinpoints* [Manuscript submitted for publication]. Department of Psychology, Appalachian State University.

Laske, M.M., & Ludwig, T.D. (2022b). *Sustainable risk identification: Manager versus employee safety processes* [Manuscript submitted for publication]. Department of Psychology, Appalachian State University.

Laske, M.M., & Ludwig, T.D. (2022c). *The effects of mandatory quotas on a behavioral safety program: A ten year case study* [Manuscript in preparation]. Department of Psychology, Appalachian State University.

Lebbon, A.R., & Sigurdsson, S.O. (2017). Behavioral perspectives on variability in human behavior as part of process safety. *Journal of Organizational Behavior Management, 37*(3–4), 261–282. https://doi.org/10.1080/01608061.2017.1340922

Lebbon, A.R., Sigurdsson, S.O., & Austin, J. (2012). Behavioral safety in the food services industry: Challenges and outcomes. *Journal of Organizational Behavior Management, 32*(1), 44–57. https://doi.org/10.1080/01608061.2011.592792

Liberty Mutual (2021). The Liberty Mutual workplace safety index 2021 (WSI 1000 R2). Boston, MA: Liberty Mutual Insurance. https://business.libertymutual.com/wp-content/uploads/2021/06/2021_WSI_1000_R2.pdf

Lindsley, O.R. (1991). From technical jargon to plain English for application. *Journal of Applied Behavior Analysis, 24*(3), 449–458. https://doi.org/10.1901/jaba.1991.24-449

Liu, H., & Tsai, Y. (2012). A fuzzy risk assessment approach for occupational hazards in the construction industry. *Safety Science, 50*(4), 1067–1078. https://doi.org/10.1016/j.ssci.2011.11.021

Ludwig, T.D. (2002). On the necessity of structure in an arbitrary world: Using concurrent schedules of reinforcement to describe response generalization. *Journal of Organizational Behavior Management*, *21*(4), 13–38. https://doi.org/10.1300/J075v21n04_03

Ludwig, T.D. (2014). The anatomy of pencil whipping. *Professional Safety*, *59*(02), 47–50.

Ludwig, T.D. (2015). Organizational behavior management: An enabler of applied behavior analysis. In T.S. Falcomata, J.E. Ringdahl, & H. Roane (eds.), *Clinical and organizational applications of applied behavior analysis* (pp. 605–625). Academic Press, Elsevier. https://doi.org/http://dx.doi.org/10.1016/B978-0-12-420249-8.00024-1

Ludwig, T.D. (2017a). Process safety: Another opportunity to translate behavior analysis into evidence-based practices of grave societal value (Editorial). *Journal of Organizational Behavior Management*, *37*(3–4), 221–223. https://doi.org/10.1080/01608061.2017.1343702

Ludwig, T.D. (2017b). Process safety behavioral systems: Behaviors interlock in complex metacontingencies. *Journal of Organizational Behavior Management*, *37*(3–4), 224–239. https://doi.org/10.1080/01608061.2017.1340921

Ludwig, T.D. (2018). *Dysfunctional practices that kill your safety culture*. Calloway Publishing.

Ludwig, T.D., Biggs, J., Wagner, S., & Geller, E.S. (2002). Using public feedback and competitive rewards to increase the safe driving of pizza deliverers. *Journal of Organizational Behavior Management*, *21*(4), 75–104. https://doi.org/10.1300/J075v21n04_06

Ludwig, T.D., & Geller, E.S. (1991). Improving the driving practices of pizza deliverers: Response generalization and moderating effects of driving history. *Journal of Applied Behavior Analysis*, *24*(1), 31–44. https://doi.org/10.1901/jaba.1991.24-31

Ludwig, T.D., & Geller, E.S. (1997). Assigned versus participative goal setting and response generalization: Managing injury control among professional pizza deliverers. *Journal of Applied Psychology*, *82*(2), 253–261. https://doi.org/10.1037/0021-9010.82.2.253

Ludwig, T.D., & Geller, E.S. (1999). Behavior change among agents of a community safety program: Pizza deliverers advocate community safety belt use. *Journal of Organizational Behavior Management*, *19*(2), 3–24. https://doi.org/10.1300/J075v19n02_02

Ludwig, T.D., & Geller, E.S. (2000). Intervening to improve the safety of delivery drivers: A systematic behavioral approach. *Journal of Organizational Behavior Management*, *19*(4), 1–124. https://doi.org/10.1300/ J075v19n04_01

Ludwig, T.D., & Geller, E.S. (2001). *Intervening to improve the safety of occupational driving*. Haworth.

Ludwig, T.D., Geller, E.S., & Clarke, S.W. (2010). The additive impact of group and individual publicly-displayed feedback: Examining individual response patterns and response generalization in a safe-driving occupational intervention. In J.K Luiselli (ed.), *Performance Psychology: Theory and Application in Industry, Sports, and Behavioral Healthcare*. A special issue of *Behavior Modification*, *34* (5), 338–366.

Ludwig, T.D., & Goomas, D.T. (2009). Real-time performance monitoring, goal-setting, and feedback for forklift drivers in a distribution centre. *Journal of Occupational and Organizational Psychology, 82*(2), 391–403. https://doi.org/10.1348/096317908X314036

Ludwig, T.D., & Houmanfar, R.A. (eds.) (2010). *Understanding complexity in organizations: Behavioral systems.* Routledge.

Ludwig, T.D., & Laske, M.M. (2019–2021). *Behavior change due to feedback: Data from grocery distribution company* [Unpublished data set].

Ludwig, T.D., Leslie, J., Granowsky, N., Acikgoz, Y., Bergman, S. (2023). *Safety analytics in three organizations suggest the probability of injury decreases after a behavioral safety observation* [Paper presentation]. Association for Behavior Analysis International 49th Annual Convention, Denver, CO, United States.

Luthans, F., & Peterson, S.J. (2003). 360-Degree feedback with systematic coaching: Empirical analysis suggests a winning combination. *Human Resource Management, 42*(3), 243–256. https://doi.org/10.1002/hrm.10083

Luthans, F., & Stajkovic, A.D. (1999). Reinforce for performance: the need to go beyond pay and even rewards. *The Academy of Management Executive, 13*(2), 49–57. https://doi.org/10.5465/ame.1999.1899548

Mahoney, M.J. (1989). Scientific psychology and radical behaviorism: Important distinctions based in scientism and objectivism. *American Psychologist, 44*(11), 1372–1377. https://doi.org/10.1037/0003-066X.44.11.1372

Malott, M.E. (2003). *Paradox of organizational change.* Context Press.

Malott, R.W. (1993). A theory of rule-governed behavior and organizational behavior management. *Journal of Organizational Behavior Management, 12*(2), 45–64. https://doi.org/10.1300/J075v12n02_03

Malott, R.W., Shimamune, S., & Malott, M.E. (1993). Rule-governed behavior and organizational behavior management: An analysis of interventions. *Journal of Organizational Behavior Management, 12*(2), 103–116. https://doi.org/10.1300/J075v12n02_09

Marquis, M.S., & Manning, W.G. (1999). Lifetime costs and compensation for injuries. *Inquiry, 36*(3), 244–254.

Matey, N., Sleiman, A., Natasi, J., Richard, E., & Gravina, N. (2021). Varying reactions to feedback and their effects on observer accuracy and feedback omission. *Journal of Applied Behavior Analysis, 54*(3), 1188–1198. https://doi.org/10.1002/jaba.840

Matthews, R. (2022). *Quantifying quality: Using quantitative methods to evaluate observation quality and its impact on incident reduction* [Unpublished master's thesis]. Appalachian State University.

Mathis, T.L. (2009). Unions and behavior-based safety: The 7 deadly sins. *EHS Today, 2*(10), 11–25.

Mawhinney, T.C. (2001). Organization-environment systems as OBM intervention context: Minding your metacontingencies. In L.J. Hayes, J.A. Austin, R. Houmanfar, & M.C. Clayton (eds.), *Organizational Change* (pp. 137–165). Context Press.

Mayer, G.R., Sulzer-Azaroff, B., & Wallace, M. (2019). *Behavior analysis for lasting change* (4th ed.). Sloan Publishing, LLC.

Mazur, G. (2003). Voice of the customer (define): QFD to define value. In Proceedings of the 57th American Quality Congress. Kansas City, MO, United States.

McSween, T.E. (1995). *The values-based safety process: Improving your safety culture with behavior-based safety*. Wiley.

McSween, T.E. (2003). *Value-based safety process: Improving your safety culture with behavior-based safety* (2nd ed.). John Wiley & Sons, Inc.

Merritt, T.A., DiGennaro Reed, F.D., & Martinez, C.E. (2019). Using the Performance Diagnostic Checklist–Human Services to identify an indicated intervention to decrease employee tardiness. *Journal of Applied Behavior Analysis*, *52*(4), 1034–1048. https://doi.org/10.1002/jaba.643

Metzgar, C. (2011). The truth wears off. *Professional Safety*, 19–21.

Miller, L.K. (2006). *Principles of everyday behavior analysis* (4th ed.). Thomson Wadsworth.

Mistikoglu, G., Gerek, I.H., Erdis, E., Usmen, P.E.M., Cakan, H., & Kazan, E.E. (2015). Decision tree analysis of construction fall accidents involving roofers. *Expert Systems with Applications*, *42*, 2256–2263. https://doi.org/10.1016/j.eswa.2014.10.009

Mollicone, D., Kan, K., Mott, C., Bartels, R., Bruneau, S., van Wollen, M., Sparrow, A.R., & Van Dongen, H.P.A. (2019). Predicting performance and safety based on driver fatigue. *Accident; Analysis and Prevention*, *126*, 142–145. https://doi.org/10.1016/j.aap.2018.03.004

Morris, E.K., Smith, N.G., & Altus, D.E. (2005). B.F. Skinner's contributions to applied behavior analysis. *Behavior Analyst*, *28*(2), 99–131. https://doi.org/10.1007/BF03392108

Myers, W.V., McSween, T.E., Medina, R.E., Rost, K., & Alvero, A.M. (2010). The implementation and maintenance of a behavioral safety process in a petroleum refinery. *Journal of Organizational Behavior Management*, *30*(4), 285–307. https://doi.org/10.1080/01608061.2010.499027

National Institute for Occupational Safety and Health (n.d.). *Hierarchy of controls*. https://www.cdc.gov/niosh/topics/hierarchy/default.html

National Institute for Occupational Safety and Health (1997). *Musculoskeletal disorders and workplace factors: A critical review of epidemiologic evidence for work-related musculoskeletal disorders of the neck, upper extremity, and low back* (Publication no. 97–141). Cincinnati OH: National Institute of Occupational Safety and Health. https://www.cdc.gov/niosh/docs/97-141/

National Institute for Occupational Safety and Health (2011). *Prevention through design: Plan for the national initiative* (NIOSH Publication No. 2011–121). https://www.cdc.gov/niosh/docs/2011-121/pdfs/2011-121.pdf?id=10.26616/NIOSHPUB201112

National Safety Council (2021). Work injury costs. https://injuryfacts.nsc.org/work/costs/work-injury-costs/

Nimmer, J.G., & Geller, E.S. (1988). Motivating safety belt use at a community hospital: an effective integration of incentive and commitment strategies. *American Journal of Community Psychology*, *16*(3), 381–394. https://doi.org/10.1007/BF00919377

Occupational Safety and Health Administration [OSHA]. (2000). *Process safety management, 3132 (reprinted)*. https://www.osha.gov/Publications/osha3132.html

Occupational Safety and Health Administration [OSHA]. (2012). *Employer safety incentive and disincentive policies and practices.* https://www.osha.gov/as/opa/whistleblowermemo.html

Occupational Safety and Health Administration [OSHA]. (2016). *Improve tracking of workplace injuries and illnesses; final rule* (81:29623–29694). https://www.osha.gov/laws-regs/federalregister/2016-05-12

Occupational Safety and Health Administration [OSHA]. (2018). *Clarification of OSHA's position on workplace safety incentive programs and post-incident drug testing under 29 C.F.R. § 1904.35(b)(1)(iv)* https://www.osha.gov/laws-regs/standardinterpretations/2018-10-11

Occupational Safety and Health Administration [OSHA]. (2019). *Using leading indicators to improve safety and health outcomes.* https://www.osha.gov/leadingindicators/docs/OSHA_Leading_Indicators.pdf

Petrock, F. (1978). Analyzing the balance of consequences for performance improvement. *Journal of Organizational Behavior Management, 1*(3), 196–205. https://doi.org/10.1300/J075v01n03_04

Predictive Solutions (2012). *Predictive analytics in workplace safety: Four 'safety truths' that reduce workplace injuries.* https://www.predictivesolutions.com/lp/four-safety-truths-reduce-workplace-injuries

Reber, R.A., & Wallin, J.A. (1983). Validation of a behavioral measure of occupational safety. *Journal of Organizational Behavior Management, 5*(2), 69–78. https://doi.org/10.1300/J075v05n02_04

Reber, R.A., & Wallin, J.A. (1984). The effects of training, goal setting, and knowledge of results on safe behavior: A component analysis. *The Academy of Management Journal, 27*(3), 544–560.

Reville, R.T., Boden, L.I., Biddle, J.E., & Mardesich, C. (2001). *An evaluation of New Mexico workers' compensation permanent partial disability and return to work.* RAND.

Rhoton, W.W. (1980). A procedure to improve compliance with coal mine safety regulations. *Journal of Organizational Behavior Management, 2*(4), 243–249. https://doi.org/10.1300/J075v02n04_01

Rummler, G. A., & Brache, A. P. (2013). *Improving performance how to manage the white space on the organization chart* (3rd ed.). Jossey-Bass.

Rudd, J.R., & Geller, E.S. (1985). A university-based incentive program to increase safety belt use: Toward cost-effective institutionalization. *Journal of Applied Behavior Analysis, 18*(3), 215–226. https://doi.org/10.1901/jaba.1985.18-215

Ryan, R.M., & Deci, E.L. (2000). Self-determination theory and the faciliation of intrinsic motivation, social development, and well-being. *American Psychologist, 55*(1), 68–78. https://doi.org/10.1037/0003–066X.55.1.68

Saal, F.E., Downey, R.G., & Lahey, M.A. (1980). Rating the ratings: Assessing the psychometric quality of rating data. *Psychological Bulletin, 88*(2), 413–428. https://doi.org/10.1037/0033-2909.88.2.413

Safety Management Group. (n.d.). *Injury cost calculator.* https://safetymanagementgroup.com/resources/injury-cost-calculator/

Sasson, J.R., & Austin, J. (2005). The effects of training, feedback, and participant involvement in behavioral safety observations on office ergonomic behavior. *Journal of Organizational Behavior Management*, 24(4), 1–30. https://doi.org/10.1300/J075v24n04_01

Schlinger, H., & Blakely, E. (1987). Function-altering effects of contingency-specifying stimuli. *The Behavior Analyst*, 10(1), 41–45. https://doi.org/10.1017/CBO9781107415324.004

Shah, R., & Ward, P.T. (2007). Defining and developing measures of lean production. *Journal of Operations Management*, 25(4), 785–805. https://doi.org/10.1016/j.jom.2007.01.019

Sidman, M. (1960). *Tactics of scientific research*. Basic Books, Inc.

Skinner, B.F. (1938). *The behavior of organisms: An experimental analysis*. Appleton-Century-Crofts.

Skinner, B.F. (1953). *Science and human behavior*. Macmillan.

Skinner, B.F. (1957). *Verbal behavior*. Appleton-Century-Crofts.

Skinner, B.F. (1966). An operant analysis of problem solving. In B. Keinmuntz (ed.), *Problem solving: Research, method, and therapy* (pp. 225–257). Wiley.

Skinner, B.F. (1969). An operant analysis of problem-solving. In *Contingencies of reinforcement: A theoretical analysis* (pp. 113–171). Prentice-Hall.

Skinner, B.F. (1974). *About Behaviorism*. Knopf.

Sleiman, A.A., Sigurjonsdottir, S., Elnes, A., Gage, N.A., & Gravina, N.E. (2020). A quantitative review of performance feedback in organizational settings (1998–2018). *Journal of Organizational Behavior Management*, 40(3–4), 303–332. https://doi.org/10.1080/01608061.2020.1823300

Smith, M.J., Anger, W.K., & Uslan, S.S. (1978). Behavior modification applied to occupational safety. *Journal of Safety Research*, 10(2), 87–88.

Smith, T.A. (1999). What's wrong with behavior-based safety? *Professional Safety, September*, 37–40.

Spigener, J., Lyon, G., & McSween, T. (2022). Behavior-based safety 2022: Today's evidence. *Journal of Organizational Behavior Management*. Advance online publication. https://doi.org/10.1080/01608061.2022.2048943

Stephens, S.D., & Ludwig, T.D. (2005). Improving anesthesia nurse compliance with universal precautions using group goals and public feedback. *Journal of Organizational Behavior Management*, 25(2), 37–71. https://doi.org/10.1300/J075v25n02_02

Stokes, T.F., & Baer, D.M. (1977). An implicit technology of generalization. *Journal of Applied Behavior Analysis*, 10(2), 349–367. https://doi.org/10.1901/jaba.1977.10-349

Stokes, T.F., & Osnes, P.G. (1989). An operant pursuit of generalization. *Behavior Therapy*, 20(3), 337–355. https://doi.org/10.1016/S0005-7894(89)80054-1

Strunin, L., & Boden, L.I. (2004). Family consequences of chronic back pain. *Social Science and Medicine*, 58(7), 1385–1393. https://doi.org/10.1016/S0277-9536(03)00333-2

Sulzer-Azaroff, B. (1978). Behavioral ecology and accident prevention. *Journal of Organizational Behavior Management*, 2, 11–44. https://doi.org/10.1300/J075v02n01

Sulzer-Azaroff, B. (1987). The modification of occupational safety behavior. *Journal of Occupational Accidents*, 9(3), 177–197. https://doi.org/10.1016/0376-6349(87)90011-3

Sulzer-Azaroff, B., & Austin, J.A. (2000). Does BBS work? *Professional Safety*, 19–24.

Sulzer-Azaroff, B., & De Santamaria, M.C. (1980). Industrial safety hazard reduction through performance feedback. *Journal of Applied Behavior Analysis*, 13(2), 287–295. https://doi.org/10.1901/jaba.1980.13-287

Sulzer-Azaroff, B., & Fellner, D.J. (1984). Searching for performance targets in the behavioral analysis of occupational health and safety: An assessment strategy. *Journal of Organizational Behavior Management*, 6(2), 53–65. https://doi.org/10.1300/J075v06n02_09

Sulzer-Azaroff, B., Loafman, B., Merante, R.J., & Hlavacek, A.C. (1990). Improving occupational safety in a large industrial plant. *Journal of Organizational Behavior Management*, 11(1), 99–120. https://doi.org/10.1300/J075v11n01_07

Swink, M., & Jacobs, B.W. (2012). Six Sigma adoption: Operating performance impacts and contextual drivers of success. *Journal of Operations Management*, 30(6), 437–453. https://doi.org/10.1016/j.jom.2012.05.001

Todd, J.T., & Morris, E.K. (1992). Case histories in the great power of steady misrepresentation. *American Psychologist*, 47(11), 1441–1453. https://doi.org/10.1037/0003-066X.47.11.1441

Townley-Cochran, D., Leaf, J.B., Taubman, M., Leaf, R., & McEachin, J. (2015). Observational learning for students diagnosed with autism: A review paper. *Review Journal of Autism and Developmental Disorders*, 2(3), 262–272. https://doi.org/10.1007/s40489-015-0050-0

Van Houten, R., Rolider, A., Nau, P.A., Friedman, R., Becker, M., Chalodovsky, I., & Scherer, M. (1985). Large-scale reductions in speeding and accidents in Canada and Israel: a behavioral ecological perspective. *Journal of Applied Behavior Analysis*, 18(1), 87–93. https://doi.org/10.1901/jaba.1985.18-87

Vaughan, M. (1989). Rule-governed behavior in behavior analysis: A theoretical and experimental history. In S.C. Hayes (ed.), *Rule-governed behavior: Cognition, contingencies, and instructional control* (pp. 97–118). Plenum Press.

Vinodkumar, M.N., & Bhasi, M. (2010). Safety management practices and safety behaviour: Assessing the mediating role of safety knowledge and motivation. *Accident Analysis and Prevention*, 42(6), 2082–2093. https://doi.org/10.1016/j.aap.2010.06.021

Vroom, V.H. (1964). *Work and motivation*. Wiley.

Waldersee, R., & Luthans, F. (1994). The impact of positive and corrective feedback on customer service performance. *Journal of Organizational Behavior*, 15(1), 83–95. https://doi.org/10.1002/job.4030150109

White, O.R. (1971). *A glossary of behavioral terminology*. Research Press Company. https://doi.org/10.1017/CBO9781107415324.004

Wilder, D.A., Lipschultz, J.L., King, A., Driscoll, S., & Sigurdsson, S. (2018). An analysis of the commonality and type of preintervention assessment procedures in the Journal of Organizational Behavior Management (2000–2015). *Journal of Organizational Behavior Management*, 38(1), 5–17. https://doi.org/10.1080/01608061.2017.1325822

INDEX

For Product Safety Concerns and Information please contact our EU
representative GPSR@taylorandfrancis.com
Taylor & Francis Verlag GmbH, Kaufingerstraße 24, 80331 München, Germany